READING ESSENTIALS AND NOTE-TAKING GUIDE

STUDENT WORKBOOK

New York, New York Columbus, Ohio Chicago, Illinois

The McGraw·Hill Companies

 Glencoe

Copyright © by The McGraw-Hill Companies, Inc. All rights reserved. Except as permitted under the United States Copyright Act, no part of this publication may be reproduced or distributed in any form or by any means, or stored in a database or retrieval system, without prior permission of the publisher.

Send all inquiries to:
Glencoe/McGraw-Hill
8787 Orion Place
Columbus, OH 43240-4027

ISBN: 978-0-07-877602-1
MHID: 0-07-877602-3

Printed in the United States of America.

6 7 8 9 10 MAL 10

Contents

To The Student vii

Chapter 1 Using Geography Skills
Section 1 Thinking Like a Geographer. 1
Section 2 The Earth in Space 4

Chapter 2 Earth's Physical Geography
Section 1 Forces Shaping the Earth. 7
Section 2 Landforms and Water Resources 10
Section 3 Climate Regions. 13
Section 4 Human-Environment Interaction 16

Chapter 3 Earth's Human and Cultural Geography
Section 1 World Population 19
Section 2 Global Cultures 22
Section 3 Resources, Technology, and World Trade 25

Chapter 4 Physical Geography of the United States and Canada
Section 1 Physical Features 28
Section 2 Climate Regions. 31

Chapter 5 History and Cultures of the United States and Canada
Section 1 History and Governments 34
Section 2 Cultures and Lifestyles. 37

Chapter 6 The United States and Canada Today
Section 1 Living in the United States and Canada Today 40
Section 2 Issues and Challenges 43

Chapter 7 Physical Geography of Latin America
Section 1 Physical Features 46
Section 2 Climate Regions. 49

Chapter 8 History and Cultures of Latin America
Section 1 History and Governments 52
Section 2 Cultures and Lifestyles. 55

Chapter 9 Latin America Today
Section 1 Mexico 58
Section 2 Central America and the Caribbean 61
Section 3 South America. 64

Chapter 10 Physical Geography of Europe
Section 1 Physical Features 67
Section 2 Climate Regions. 70

Contents

Chapter 11 History and Cultures of Europe

Section 1 History and Governments 73
Section 2 Cultures and Lifestyles 76

Chapter 12 Europe Today

Section 1 Northern Europe 79
Section 2 Europe's Heartland 82
Section 3 Southern Europe 85
Section 4 Eastern Europe 88

Chapter 13 Physical Geography of Russia

Section 1 Physical Features 91
Section 2 Climate and the Environment 94

Chapter 14 History and Cultures of Russia

Section 1 History and Governments 97
Section 2 Cultures and Lifestyles 100

Chapter 15 Russia Today

Section 1 A Changing Russia 103
Section 2 Issues and Challenges 106

Chapter 16 Physical Geography of North Africa, Southwest Asia, and Central Asia

Section 1 Physical Features 109
Section 2 Climate Regions 112

Chapter 17 History and Cultures of North Africa, Southwest Asia, and Central Asia

Section 1 History and Religion 115
Section 2 Cultures and Lifestyles 118

Chapter 18 North Africa, Southwest Asia, and Central Asia Today

Section 1 North Africa 121
Section 2 Southwest Asia 124
Section 3 Central Asia 127

Chapter 19 Physical Geography of Africa South of the Sahara

Section 1 Physical Features 130
Section 2 Climate Regions 133

Chapter 20 History and Cultures of Africa South of the Sahara

Section 1 History and Governments 136
Section 2 Cultures and Lifestyles 139

iv

Contents

Chapter 21 Africa South of the Sahara Today
Section 1 West Africa 142
Section 2 Central and East Africa. 145
Section 3 Southern Africa 148

Chapter 22 Physical Geography of South Asia
Section 1 Physical Features 151
Section 2 Climate Regions. 154

Chapter 23 History and Cultures of South Asia
Section 1 History and Governments. 157
Section 2 Cultures and Lifestyles. 160

Chapter 24 South Asia Today
Section 1 India 163
Section 2 Muslim Nations. 166
Section 3 Mountain Kingdoms, Island Republics 169

Chapter 25 Physical Geography of East Asia and Southeast Asia
Section 1 Physical Features 172
Section 2 Climate Regions. 175

Chapter 26 History and Cultures of East Asia and Southeast Asia
Section 1 History and Governments. 178
Section 2 Cultures and Lifestyles. 181

Chapter 27 East Asia and Southeast Asia Today
Section 1 China. 184
Section 2 Japan 187
Section 3 The Koreas. 190
Section 4 Southeast Asia. 193

Chapter 28 Physical Geography of Australia, Oceania, and Antarctica
Section 1 Physical Features 196
Section 2 Climate Regions. 199

Chapter 29 History and Cultures of Australia, Oceania, and Antarctica
Section 1 History and Governments. 202
Section 2 Cultures and Lifestyles. 205

Chapter 30 Australia, Oceania, and Antarctica Today
Section 1 Australia and New Zealand. 208
Section 2 Oceania 211
Section 3 Antarctica 214

To The Student

Taking good notes helps you become more successful in school. Using this book helps you remember and understand what you read. You can use this *Reading Essentials and Note-Taking Guide* to improve your test scores. Some key parts of this booklet are described below.

The Importance of Graphic Organizers

First, many graphic organizers appear in this *Reading Essentials and Note-Taking Guide*. Graphic organizers allow you to see important information in a visual way. Graphic organizers also help you understand and summarize information, as well as remember the content.

The Cornell Note-Taking System

Second, you will see that the pages in the *Reading Essentials and Note-Taking Guide* are arranged in two columns. This two-column format is based on the **Cornell Note-Taking System,** developed at Cornell University. The large column on the right side of the page contains the essential information from each section of your textbook, *Exploring Our World.*

The column on the left side of the page includes a number of note-taking prompts. In this column, you will perform various activities that will help you focus on the important information in the lesson. You will use recognized reading strategies to improve your reading-for-information skills.

Vocabulary Development

Third, you will notice that vocabulary words are bolded throughout the *Reading Essentials and Note-Taking Guide.* Take special note of these words. You are more likely to be successful in school when you have vocabulary knowledge. When researchers study successful students, they find that as students acquire vocabulary knowledge, their ability to learn improves.

Writing Prompts and Note-Taking

Finally, a number of writing exercises are included in this *Reading Essentials and Note-Taking Guide.* You will see that many of the note-taking exercises ask you to practice the critical-thinking skills that good readers use. For example, good readers *make connections* between their lives and the text. They also *summarize* the information that is presented and *make inferences* or *draw conclusions* about the facts and ideas. At the end of each section, you will be asked to respond to two short-answer questions and one essay. The essays prompt you to use one of four writing styles: informative, descriptive, persuasive, or expository.

The information and strategies contained within the *Reading Essentials and Note-Taking Guide* will help you better understand the concepts and ideas discussed in your social studies class. They also will provide you with skills you can use throughout your life.

Chapter 1, Section 1 (Pages 14–17)
Thinking Like a Geographer

Big Idea

Geography is used to interpret the past, understand the present, and plan for the future. As you read, complete the chart below by identifying two examples for each topic.

Themes of Geography
1. location
2. movement
Types of Geography
1. Physical Geography
2. Human Geography
Geographers' Tools
1. GIS
2. GPS

Notes | Read to Learn

The Five Themes of Geography (page 15)

Explaining

Explain the difference between absolute location and relative location.

absolute location: refers to the exact spot on Earth. relative location: where a feature is in relation to the features around it.

Geography is the study of Earth and its people. Scientists who do this work are geographers. They use five main themes to describe people and places. The five themes of geography are location, place, human-environment interaction, movement, and regions.

The position of a place on Earth's surface is its *location*, which can be described in two ways. **Absolute location** refers to the exact spot on Earth where a place or feature is found. **Relative location** explains where a feature is in relation to the features around it.

Place refers to the characteristics of a location that make it unique. One way to define a place is by its physical features—landforms, plants, animals, and weather patterns. A place also can be defined by its human characteristics, such as its language.

The **environment** is one's natural surroundings. *Human-environment interaction* explores how people affect, and are affected by, their environment. People affect the environment by changing it to meet their needs. People, in turn, are influenced by environmental factors they cannot control, such as temperature and natural disasters.

The Five Themes of Geography (continued)

Applying

What region do you live in?

Mid-Atlantic

Movement explores how and why people, ideas, and goods move from one place to another. For example, people might move to flee from a country that is at war. Movement causes cultural change.

Regions are areas of the Earth's surface that have features in common. These features may be land, natural resources, or population. For example, the Rocky Mountain region of the United States is known for ranching and mining.

A Geographer's Tools (pages 16–17)

Sequencing

Write down the four long periods of history from the earliest to the most recent.

1. Prehistory
2. Ancient History
3. Middle Ages
4. Modern History

Stating

What types of information do satellites provide to mapmakers?

They can measure temperature and pollution.

Types of Geography

Geographers study Earth's physical and human features. Physical geographers study land areas, bodies of water, plant life, and other physical features. They also examine natural resources and the ways people use them.

Human geographers study people and their activities. They examine religions, languages, and ways of life. Human geographers can focus on a specific location or look at broader areas. They often make comparisons between different places.

Places in Time

Geographers study history to learn about changes that have occurred over time. History is divided into blocks of time called periods. A **decade** is a period of 10 years. A **century** is a period of 100 years. A **millennium** is a period of 1,000 years.

In Western society, history is commonly grouped into four long periods. Prehistory is the time before writing was developed. This period ended about 5,500 years ago. The next period, which lasted until 1,500 years ago, is Ancient History. That period was followed by the Middle Ages, which lasted about 1,000 years. Modern History is the period from about 500 years ago through the present.

Map Systems

Geographers use maps to study different types of information about a place. Some maps are created from information collected by satellites that circle the Earth. For example, satellites provide photographs and can measure changing temperatures and pollution. A specific group of satellites makes up the **Global Positioning System (GPS)**. This system uses radio signals to record the precise location of every place on Earth. GPS devices are installed in cars and trucks and used by hikers so they do not get lost.

 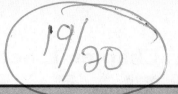

19/20

A Geographer's Tools (continued)

Differentiating

What is the difference between GPS and GIS?

<u>GIS: is computer hardware and software that collect geographic data and display it on a screen. what about GPS?</u>

Geographic Information Systems (GIS) are computer hardware and software that collect geographic data and display the data on a screen. GIS provides more detailed information that does not usually appear on maps, such as types of soil and water quality.

Careers in Geography

Careers for geographers exist at all levels of government and in private businesses. Governments hire geographers to help determine how land and resources are best put to use. Geographers also study population trends and help plan cities. Businesses hire geographers to locate resources, decide where to set up new offices, and provide information about places and cultures that companies deal with.

Section Wrap-Up

Answer these questions to check your understanding of the entire section.

1. **Distinguishing** Explain the difference between *place* and *location*.

   <u>Place refers to the characteristics of a location that make it unique. Location is the position of a place on the Earth's surface.</u>

2. **Making Connections** Complete this chart with examples of what physical geographers and human geographers study.

Physical Geographers	Human Geographers
• land areas • bodies of water • plant life	• study people • religion • language

 *Think about the different career options for geographers. On a separate sheet of paper, write a paragraph about a job in geography that you might enjoy.*

Chapter 1, Section 1

Chapter 1, Section 2 (Pages 34–38)
The Earth in Space

36/40

Big Idea

Physical processes shape Earth's surface. As you read, complete the diagram below. Explain the effects of latitude on Earth's temperature.

Tropics	→	The Tropics tend to be warm.

High Latitudes	→	These polar regions are always cool or cold.

Notes | Read to Learn

The Solar System (pages 35–36)

Naming

Name the eight major planets in our solar system.

1. Mercury
2. Venus
3. Earth
4. Mars
5. Jupiter
6. Saturn
7. Neptune
8. Uranus

Eight major planets, including Earth, revolve around the sun. Thousands of smaller bodies also circle the sun. All of these, together with the sun, form our **solar system.**

Major Planets

The eight major planets differ from each other in size and form. The four inner planets closest to the sun are Mercury, Venus, Earth, and Mars. They are relatively small and solid.

Jupiter, Saturn, Uranus, and Neptune are the four outer planets. They are larger and formed mostly or entirely of gases. Pluto, once considered a major planet, is now classified as a minor planet.

Each planet follows its own **orbit,** or path, around the sun. Some orbits are almost circular, whereas others are oval shaped. The lengths of the orbits also vary, from 88 days for Mercury to 165 years for Neptune.

Earth's Movement

Earth makes a **revolution,** or complete circuit, around the sun every 365¼ days. This time period is defined as one year.

The Solar System (continued)

Explaining

Why do people not feel Earth move?

Because the atmosphere moves with it.

Every four years is a **leap year,** when the extra fourths of a day are combined and added to the calendar as February 29.

Earth **rotates,** or spins, on its axis as it orbits the sun. The **axis** is an imaginary line that passes through the center of Earth from the North Pole to the South Pole. Earth rotates in an easterly direction. It takes 24 hours for Earth to complete a single rotation. As it rotates, different parts of Earth are in sunlight, which is defined as daytime. Those parts facing away from the sun are in darkness and experience night. A layer of oxygen and gases, called the **atmosphere,** surrounds Earth. As Earth rotates, the atmosphere moves with it, so people do not feel Earth moving.

Sun and Seasons (pages 37–38)

Summarizing

What causes Earth to experience changing seasons?

The tilt of Earth on its axis.

Earth is tilted 23½ degrees on its axis. This tilt causes seasons to change as Earth orbits the sun. The tilt determines whether or not an area will receive direct rays from the sun. When a hemisphere receives direct rays, it has summer. When a hemisphere receives indirect, or slanted, rays, it experiences the cold of winter.

Solstices and Equinoxes

The North Pole is tilted toward the sun on or about June 21, and the sun is directly over the Tropic of Cancer (23½°N latitude). This day is called the **summer solstice.** In the Northern Hemisphere, June 21 has the most hours of sunlight and marks the beginning of summer. On the same day, the Southern Hemisphere has the fewest hours of sunlight, and winter begins there.

Paraphrasing

Fill in the blanks.

The day with the most hours **of sunlight is the beginning of summer. The first day of winter is the day with the** least hours **of sunlight.**

Six months later, on or about December 22, the North Pole is tilted away from the sun and the sun's direct rays hit the Tropic of Capricorn (23½°S latitude). This is the **winter solstice** for the Northern Hemisphere—the day with the fewest hours of sunlight and the beginning of winter. It is the first day of summer in the Southern Hemisphere, however.

Midway between the two solstices are the **equinoxes,** when day and night are of equal length in both hemispheres. The equinoxes mark the beginning of spring and fall. The spring equinox occurs on or about March 21, and the fall equinox occurs around September 23. On both days, the noon sun shines directly over the Equator.

Sun and Seasons (continued)

Identifying

Identify each latitude region.

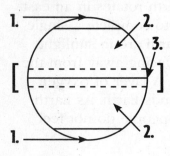

1. Polar region
2. the Tropics
3. Equator

Effects of Latitude

The **Tropics** is the low-latitude region near the Equator between the Tropic of Cancer and the Tropic of Capricorn. The sun's rays hit this area directly year-round, so temperatures in the Tropics tend to be warm. In contrast, the sun's rays are always indirect at the high-latitude areas near the North and South Poles. These polar regions are always cool or cold. The areas between the Tropics and the polar regions are called the midlatitudes. Temperatures, weather, and the seasons vary widely in these areas.

Section Wrap-Up

Answer these questions to check your understanding of the entire section.

1. **Explaining** How long is Earth's orbit? How long is Earth's rotation?

   Earth's orbit is about 365 days long.
   Earth's rotation is about 24 hours long.

2. **Organizing** Complete this chart of seasons in the Northern Hemisphere. Add the approximate date when each season begins and the name of the first day for each season.

Season	Date Season Begins	Name of First Day
Winter	December 22	Winter Solstice
Spring	March 21st	Spring Equinox
Summer	June 21st	Summer Solstice
Fall	September 23	Fall Equinox

Expository Writing

On a separate sheet of paper, explain why day and night are not always the same length throughout the year.

Because of daylight savings time.
Did you read and understand the solstices and equinoxes?

Chapter 2, Section 1 (Pages 44–48)
Forces Shaping the Earth

Big Idea

Physical processes shape the Earth's surface. As you read, complete this diagram by listing the forces shaping Earth and the effects of each.

Forces		Effects
?	→	?

Notes | Read to Learn

Inside the Earth (page 45)

Labeling

As you read, write the name of each layer of Earth beside the correct number.

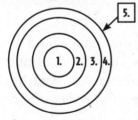

1. inner core
2. outer core
3. mantle
4. crust
5. continents

The ground feels solid when you walk on it. But Earth is not a large, solid rock. Earth has several layers, like a melon or a baseball. The three main layers of the Earth are the core, the mantle, and the crust.

Scientists divide the **core** into the inner core and the outer core. At the center of the Earth is a solid inner core of iron and nickel. It is about 3,200 miles below the surface. Scientists think the inner core is under great pressure. The next layer, the outer core, is so hot that the metal has melted into a liquid.

The **mantle** surrounds the core. It is a layer of hot, thick rock. The section of the mantle nearest the core—the inner mantle—is solid. However, the rock in the outer mantle can be moved, shaped, and even melted. The melted rock is called **magma.** It flows to the surface of the Earth when a volcano erupts. Magma is called lava when it reaches the surface.

The thin, outside layer of the Earth is the **crust.** It is a rocky shell that forms the Earth's surface. The crust includes the ocean floors. It also includes seven large land areas known as **continents.** The continents are North America, South America, Europe, Asia, Africa, Australia, and Antarctica.

Read to Learn

Shaping the Earth's Surface (pages 46–48)

Identifying

As you read, identify two forces inside the Earth and two forces outside the Earth that change the appearance of Earth.

Inside the Earth:

Volcanoes

Outside the Earth:

Earthquakes

Explaining

How and why do the continents move?

Each continent sits on one or more large bases, called plates. As these plates move, the continents move.

The Earth's crust moves and changes over time. Old mountains are worn down, and new mountains are pushed higher. The continents also move.

Plate Movements

The theory of **plate tectonics** explains how the continents were formed and why they move. Each continent sits on one or more large bases, called plates. As these plates move, the continents move. This movement is called continental drift. Earth's plates move very slowly, about 1 to 7 inches per year. Some 200 million years ago, all the continents were joined together in a huge landmass that scientists named Pangaea.

When Plates Meet

The movements of the Earth's plates shape the surface of the Earth. The plates may pull away from each other. This movement usually occurs in ocean areas, although it is also happening in East Africa. Plates also can run into each other, or collide. When two continental plates collide, they push against each other with tremendous force. The land where the plates meet is pushed up, which forms mountains.

Collisions of continental and oceanic plates produce a different result. The ocean plate is thinner. It slides underneath the continental plate. As the oceanic plate is forced down, magma in the Earth's mantle builds up. Volcanic mountains form as the magma erupts and hardens.

Earthquakes are sudden, violent movements in the Earth's crust. Many earthquakes happen in areas where the collision of oceanic plates and continental plates makes the Earth's crust unstable.

Earth's plates also can move alongside each other. This movement makes cracks in the Earth's crust called **faults**. Movements along faults result in sudden shifts that cause earthquakes.

Weathering

Volcanoes and earthquakes may cause immediate and drastic changes to Earth's surface. But other factors continue to change the landforms of the Earth at a slower pace. **Weathering** occurs when water, ice, chemicals, and even the roots of plants break rocks apart into smaller pieces.

Erosion

Weathered rock is then moved by water, wind, and ice in a process called **erosion**. Rivers and streams cut through mountains and hills. Ocean waves pound at rocks on the coast. Wind carries small bits of rock, which wear down larger rocks.

Section Wrap-Up

Answer these questions to check your understanding of the entire section.

1. **Identifying** What makes up the Earth's crust?

   The crust includes the ocean floors. It also includes the continents.

2. **Applying** What is plate tectonics, and what does it have to do with the Earth's shape?

   Each continent sits on one or more large bases, also called tectonic plates. The Earth's crust moves and changes over time. Volcanoes and mountains can form because of the plates.

Descriptive Writing

In the space provided, write an article for a science magazine describing how a valley was formed from weathering or erosion.

How Valleys Form   by: Maheen Aisha Kamal

The first thing that must take place is weathering. Weathering occurs when water, ice, chemicals, and even the roots of plants break rocks apart into smaller pieces. Weathered rock is then moved by water, wind, and ice in a process called erosion. Rivers and streams cut through mountains and hills. This forms a valley.

Chapter 2, Section 1

9

Chapter 2, Section 2 (Pages 49–54)
Landforms and Water Resources

39/40

Big Idea

Geographic factors influence where people settle. As you read, complete this diagram by identifying the various bodies of water that can be found on Earth's surface.

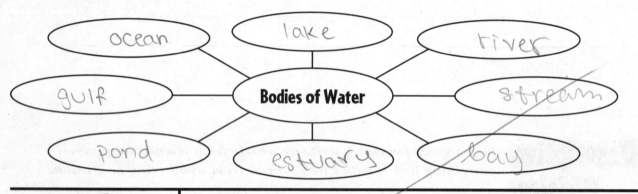

Bodies of Water: ocean, lake, river, gulf, stream, pond, estuary, bay

Notes | Read to Learn

Types of Landforms (pages 50–52)

Specifying

Which landforms are on continents, and which are on ocean floors?

On continents:
mountains
hills
flatlands
_____

On ocean floors:
trenches
continental
shelf

Earth has many landforms, from mountains to lowlands. Some of the landforms can be found both on the continents and on the ocean floors.

On Land

Mountains—huge towers of rock—are the highest landforms. Hills are lower and more rounded than mountains. The long stretches of land found between mountains or hills are called valleys. There are two types of flatlands. Plains are flatlands in low-lying areas, frequently found along coasts and river valleys. Flatlands at higher elevations are called plateaus.

An isthmus is a narrow strip of land that connects two larger landmasses. It has water on two sides. A peninsula is a piece of land that is connected to a larger landmass on one side but has water on the other three sides. An island is a body of land smaller than a continent and is completely surrounded by water.

Under the Oceans

Off the coast of any continent is an underwater plateau called a **continental shelf**. Continental shelves stretch for

Types of Landforms (continued)

several miles until finally dropping sharply to the ocean floor. On the ocean floor, tall mountains rise along the edges of the ocean plates. Tectonic activity also makes deep cuts in the ocean floor called **trenches.**

Humans and Landforms

People live on all types of landforms. Climate—the average temperature and rainfall—is one factor that helps people decide where to live. The availability of resources is another factor.

The Water Planet (pages 52–54)

Name the bodies of water that are salt water and those that are freshwater.

Salt water:

seas
bays
gulf
_____
_____

Freshwater:

groundwater
aquifers
_____

Nearly 70 percent of Earth's surface is covered with water. Earth is covered with so much water that it is sometimes called the "water planet." Water exists in many forms. It can be found in liquid form, such as streams, rivers, lakes, seas, and oceans. The atmosphere holds water in the form of a gas known as vapor. Glaciers and ice sheets are made up of water that has been frozen solid.

Salt Water

All of the oceans on Earth are actually one large body of salt water. Most of the water on Earth—almost 96 percent—is salt water. Oceans flow into smaller areas—seas, bays, and gulfs—that are somewhat enclosed by land. These, too, hold salt water, and they are linked to oceans by narrow bodies of water called straits or channels.

Freshwater

Only 4 percent of the water on Earth is freshwater, and much of this is frozen in polar ice. **Groundwater** is also freshwater. It filters through the soil, often flowing through underground layers of rock called **aquifers.** Many communities get their freshwater by pumping it from aquifers.

Freshwater is also found in lakes and rivers. A lake is a large inland body of water. Most, but not all, lakes are freshwater. Rivers are long, flowing bodies of water that begin at a source and end at a mouth. The mouth is where a river empties into another body of water, such as an ocean or a lake.

Large rivers often have tributaries, or smaller rivers and streams that feed into them. Many rivers also form deltas at their mouths. A delta is an area where a river separates into streams that flow toward the sea. Soil carried by the river settles in the delta, eventually building up the land there.

Notes | Read to Learn

The Water Planet (continued)

Listing

List the four steps in the water cycle.

1. evaporation
2. condensation
3. precipitation
4. collection

The Water Cycle

The total amount of water on Earth does not change, but it does move from place to place. Water circulates in a process called the **water cycle,** moving from the oceans, to the air, to the ground, and back to the oceans.

The sun's heat begins the water cycle by evaporating the water on the Earth's surface. **Evaporation** changes water from a liquid to water vapor. Water vapor rises from oceans and other bodies of water and spreads throughout the atmosphere.

When the air temperature cools, **condensation** takes place, meaning the water changes back into a liquid. Droplets of water in clouds fall to the ground as **precipitation.** This can take the form of rain, snow, sleet, or hail. Water then collects on the ground and in rivers, lakes, and oceans. This last step of the water cycle is known as **collection.**

Section Wrap-Up

Answer these questions to check your understanding of the entire section.

1. **Specifying** How does water help you determine whether a landform is an isthmus, a peninsula, or an island?

   It depends on whether the water is on all sides most of the sides

2. **Illustrating** Draw a diagram to show how the water cycle works. Be sure to label the various parts.

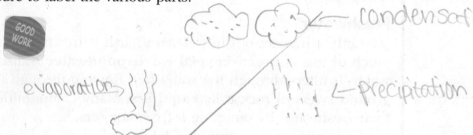

Expository Writing

On a separate sheet of paper, write a paragraph explaining what you think the advantages and disadvantages are of living on a mountain or living in a valley.

You get to see lots of cool animals on mountains. But it is hard to grow stuff on mountains. There is water on some valleys, so that is good. If water floods the valley though many things will be destroyed.

Chapter 2, Section 3 (Pages 55–61)
Climate Regions

Big Idea

Geographers organize the Earth into regions that share common characteristics. As you read, complete this diagram by identifying the effects of El Niño and La Niña.

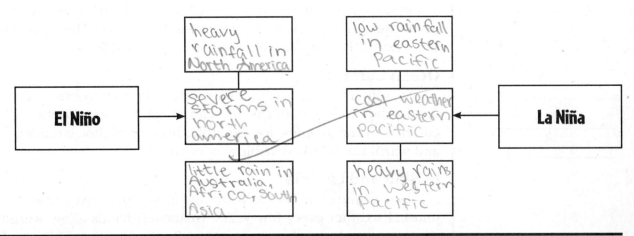

El Niño → heavy rainfall in North America; severe storms in north america; little rain in Australia, Africa, South Asia

La Niña → low rainfall in eastern Pacific; cool weather in eastern Pacific; heavy rains in western Pacific

Notes / Read to Learn

Effects on Climate (pages 56–58)

Listing

List three main things that affect climate.

1. Sun
2. Winds
3. Ocean Currents

Geographers study weather and climate. **Weather** is the short-term changes in temperature, wind direction and speed, and air moisture in a particular location. **Climate,** in contrast, is the long-term, predictable patterns of weather for a region.

The Sun

The sun's heat directly affects Earth's climate. Earth is not heated evenly by the sun. The Tropics receive more of the sun's heat energy than do the Poles. The movement of air and water, however, helps spread the sun's heat around the globe.

Winds

Movements of air are called winds. Although winds can blow in any direction, major wind systems follow patterns called **prevailing winds.** For example, warm prevailing winds in the Tropics move north and south toward the Poles. Cold polar winds move toward the Equator.

Winds curve around the Earth as it rotates. Winds that blow from east to west between the Equator and the Tropics are called trade winds. Winds that move from west to east between the Tropics and 60° north latitude are known as the westerlies.

Effects on Climate (continued)

Finding the Main Idea

Write the main idea of this subsection.

Storms_____

_____

_____

_____

Storms occur when moist, warm air rises suddenly and meets dry, cold air. Some storms include thunder and lightning as well as heavy rain. Storms may also include tornadoes—violent, funnel-shaped windstorms—or turn into blizzards.

Even more destructive storms are hurricanes and typhoons. These powerful storms begin in the Tropics and can grow to 300 miles across before striking coastal areas. Hurricanes occur in the western Atlantic and eastern Pacific Oceans. Hurricanes in the western Pacific Ocean are called typhoons.

Ocean Currents

Steadily flowing ocean streams are called **currents.** Like winds, they follow patterns and can affect climate. Ocean currents in the Tropics, for example, flow to higher latitudes and warm the winds that blow over them.

El Niño and La Niña

Two changes in Pacific wind and water patterns cause unusual weather every few years. Weakened winds allow warmer waters to reach the western coast of South America. This condition, called **El Niño,** results in heavy rains and flooding there and severe storms in North America. Yet little rain falls in Australia, southern Asia, and Africa.

La Niña, in contrast, brings low rainfall and cooler weather to the eastern Pacific. But in the western Pacific, La Niña produces storms with heavy rains, and typhoons may develop.

Landforms and Climate (pages 58–59)

Determining Cause and Effect

Why is the temperature on the tops of mountains very cold?

Because the
air is thinner
_____
_____

The shape of the land, as well as its nearness to water, affects the climate. Some landforms cause **local winds,** which are wind patterns found in a small area. Some local winds form near water. These winds occur because land warms and cools more quickly than water.

Mountains, Temperature, and Rainfall

Mountains can affect the temperature and rainfall of a local area. The air is thinner at higher elevations. The temperature at the tops of mountains is often very cold because the thin air cannot hold heat well.

A **rain shadow** occurs when mountains block rain from reaching interior regions. The side of a mountain facing the wind—the windward side—can get a large amount of rainfall. The land on the other side of the mountain—the leeward side—can be very dry. Deserts can develop on the leeward side.

Climate Zones (pages 59–61)

Listing

List the five major climate zones on Earth.

1. tropical
2. dry
3. midlatitude
4. high latitude
5. highland

A **climate zone** is an area with a particular pattern of temperature and precipitation. Areas in different parts of the world can have the same climate zone, meaning they have a similar climate. They also will have similar vegetation. Climate zones include **biomes**. These are areas in which certain kinds of plants and animals have adapted to the climate. Examples of biomes are rain forest, desert, grassland, and tundra.

Major Climates

Earth has five major climate zones—tropical, dry, midlatitude, high latitude, and highland. Four of these zones also have subcategories. For example, the dry climate zone is further divided into steppe and desert subcategories. These mostly dry climates vary slightly in rainfall and temperature.

Urban Climates

Large cities, or urban areas, often have a climate that is different from their surrounding areas. **Urban climates** have higher temperatures and unusual local winds. Paved streets and stone buildings soak up and release more of the sun's heat. The different heat patterns cause winds to blow into cities from several directions.

Section Wrap-Up

Answer these questions to check your understanding of the entire section.

1. **Determining Cause and Effect** How can winds affect the weather?
   Storms can occur when moist, warm air rises suddenly and meets dry, cold air. Winds can also warm or cool places.

2. **Specifying** What three factors define a particular climate zone?
   temperature, precipitation, vegetation

On a separate sheet of paper, write a description of the trade winds used by early explorers. Include in your account an additional description of the "horse latitudes" and why they were called that.

Chapter 2, Section 3

Chapter 2, Section 4 (Pages 63–66)
Human-Environment Interaction

Big Idea

All living things are dependent upon one another and their surroundings for survival. As you read, complete this chart by identifying four environmental problems and what people are doing to solve them.

Problem	Solution

Notes | Read to Learn

The Atmosphere (page 64)

Listing

As you read, list five negative effects of air pollution.

1. _____
2. _____
3. _____
4. _____
5. _____

People burn oil, coal, and gas for electricity, to power factories, and to move cars. Unfortunately, burning oil, coal, and gas causes air pollution.

Air Pollution

Air pollution can lead to smog. **Smog** is a thick haze of smoke and chemicals. People may have breathing problems when smog settles above cities.

Chemicals in air pollution can combine with precipitation to form **acid rain.** Acid rain kills fish, eats away at the surfaces of buildings, and destroys trees and other plant life.

Ozone is found in the atmosphere. It provides a shield against damaging rays from the sun. Some human-made chemicals destroy the ozone layer. This may result in more humans getting skin cancer.

The Greenhouse Effect

The **greenhouse effect** occurs when gases in the atmosphere trap the sun's heat. Because this heat is trapped, Earth remains warm, allowing living things to survive. Without the greenhouse effect, Earth would be too cold for most life-forms.

The Atmosphere (continued)

Differentiating

How is the greenhouse effect positive and negative?

_____
_____
_____

Some scientists believe air pollution strengthens the greenhouse effect. They claim that the burning of coal, oil, and natural gas traps the sun's heat near the Earth's surface and raises the planet's overall temperature. Such global warming could cause climate changes, melt polar ice, and result in flooding of coastal areas.

Many nations are trying to reduce global warming. Their focus is on using oil and coal more efficiently and cleanly. They also are looking at forms of energy that do not pollute. These include wind and solar power.

The Lithosphere (page 65)

Finding the Main Idea

Write down the main idea of the passage.

_____
_____
_____
_____
_____
_____

The Earth's crust is also known as the lithosphere. It includes all land above and below the oceans. Activities such as farming and cutting down trees can harm the lithosphere.

Topsoil is an important part of the lithosphere. It can be carried away by wind or water. Farming also puts topsoil at risk. However, farmers can limit the loss of topsoil in several ways. Instead of plowing straight rows, they can plow along the curves of the land. This is called contour plowing. Farmers also use **crop rotation,** which means they rotate or change what is planted from year to year. In addition, farmers plant grasses in fields without crops. The grasses hold the topsoil in place.

Another danger to topsoil is **deforestation,** or cutting down forests without replanting them. Tree roots hold soil in place. When trees are cut, the topsoil may be eroded. Trees are important for other reasons too. They support the water cycle and replace oxygen in the atmosphere. Forests are also home to many plants and animals.

The Hydrosphere and Biosphere (page 66)

Defining

Define the biosphere and the hydrosphere.

_____
_____

Surface water and groundwater make up the Earth's hydrosphere. Water is necessary for human life, but the amount of freshwater is limited. Therefore, people should practice **conservation.** This means resources such as water should be used carefully so they are not wasted.

Irrigation is a way farmers collect water and then use it on their crops. Irrigation often wastes water. Much of the water evaporates or soaks into the ground before it even gets to the crops.

Chapter 2, Section 4

Notes | Read to Learn

The Hydrosphere and Biosphere (continued)

Pollution harms the hydrosphere. **Pesticides** are chemicals that farmers use to kill insects. Strong pesticides and other chemicals sometimes spill into clean water. Then the water supply is threatened.

The biosphere is the "living" part of the planet—all the plants and animals. The biosphere is divided into many ecosystems. An **ecosystem** is a particular environment shared by certain plants and animals that need one another to survive.

Earth's **biodiversity**, or the variety of plants and animals, is getting smaller. Human activities have led to fewer types of plants and animals in some ecosystems.

Section Wrap-Up

Answer these questions to check your understanding of the entire section.

1. **Identifying** What human activities contribute to global warming?

   _____
   _____
   _____

2. **Determining Cause and Effect** How does contour plowing prevent the loss of topsoil?

   _____
   _____
   _____

Informative Writing

Create an outline for a presentation about the ways in which the hydrosphere is threatened.

_____
_____
_____
_____
_____
_____

Chapter 3, Section 1 (Pages 72–76)
World Population

Big Idea

The world's population is increasing, yet people live on only a small part of the Earth's surface. As you read, complete the diagram below to show the causes and effects of global migration.

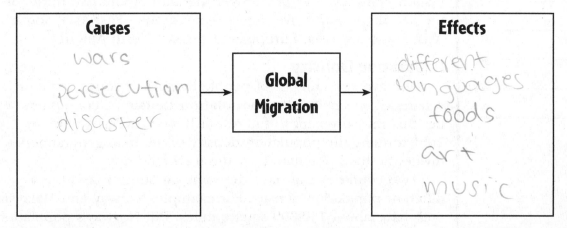

Causes: wars, persecution, disaster

Global Migration

Effects: different languages, foods, art, music

Notes — Read to Learn

Population Growth (page 73)

Listing

List two challenges created by rapid population growth.

1. food shortages
2. house shortages

The world's population has grown quickly in the past 200 years. One billion people lived on Earth around 1800. Today the population has risen to more than 6 billion. This rapid growth creates challenges for many countries.

One reason the population has grown so rapidly is because the death rate has decreased. The **death rate** is the number of deaths per year for every 1,000 people. The death rate has gone down because of better health care, improved living conditions, and an increase in the food supply.

A second cause of the population growth is high birthrates in Asia, Africa, and Latin America. The **birthrate** is the number of births per year for every 1,000 people.

Rapidly growing populations often experience food shortages. New **technology,** such as better irrigation methods and hardier crops, has helped increase the supply of food. But **famine,** or a severe lack of food, can occur as a result of warfare or crop failures.

Rapidly growing populations may experience shortages of clean water and housing as well. They also may lack hospital services and good schools.

 **Notes** | **Read to Learn**

Where People Live (page 74)

Identifying

In what regions are most of the world's people clustered?

East Asia
South Asia
Southeast Asia
Europe
Eastern North America

Only 30 percent of Earth is covered by land. Further, only half of this land can support large numbers of humans.

Population Distribution
People tend to live in areas that have fertile soil, mild climates, natural resources, and water. Therefore, the world's population is not evenly spread over the land. About two-thirds of the people on Earth live in just five regions—East Asia, South Asia, Southeast Asia, Europe, and eastern North America.

Population Density
The average number of people living in a square mile or square kilometer is called **population density.** Geographers use this measurement to figure out how crowded an area is. To determine the population density of an area, geographers divide the total population by the total land area.

Two countries can have the same amount of land but a different population density. For example, Norway and Malaysia both have about 130,000 square miles. But Norway's population density is 40 people per square mile, whereas Malaysia's population density is 205 people per square mile.

Population Movement (pages 75–76)

Defining

What is urbanization?

the growth
of cities

Differentiating

How are refugees different from emigrants?

Refugees are
forced to flee
unlike emigrants

For thousands of years, people have moved from one place to another. People and groups continue to move today.

Types of Migration
When people move from place to place within a country, it is called internal migration. These people often move from a farm or rural area into a city. Such movement leads to **urbanization,** or the growth of cities.

Movement from one country to another is called international migration. When people **emigrate,** they leave their birth country. People who emigrate are called *emigrants* in their home country and *immigrants* in their new country.

Reasons People Move
Sometimes people are "pushed" to emigrate because of negative factors in their home country. Perhaps there is a shortage of jobs or a lack of good farmland. At other times, people are "pulled," or attracted, to the new country by such things as job opportunities. **Refugees** are people who are forced to flee their country in order to escape wars, persecution, or natural disasters.

Population Movement (continued)

Impact of Migration

Migration can have positive and negative effects on both the home country and the new country. Emigration might ease overcrowding, but families can be divided. If skilled workers leave a country, the economy can suffer. Immigration can enrich the culture of the new country with different music, art, foods, and traditions. However, some native-born citizens may resent the immigrants. This can lead to unjust treatment and even violence.

Section Wrap-Up

Answer these questions to check your understanding of the entire section.

1. **Explaining** What are two causes of population growth? Explain each.

   Better health care and plentiful food supply.

2. **Speculating** Why do geographers study the population density of a country in addition to its total population?

   Because two countries can have the same amount of land but a different population density.

In the space provided, write a paragraph explaining how your community might be affected if a large number of people emigrated from the area or immigrated to it.

If a large number of people moved to my community they would probably have to build more houses and buildings. If a large number of people moved out, there would probably be less teachers. Then, more people would move.

Chapter 3, Section 1

Chapter 3, Section 2 (Pages 82–89)
Global Cultures

Big Idea

The world is made up of different cultures with some common traits. As you read, complete the diagram below by identifying the elements of culture.

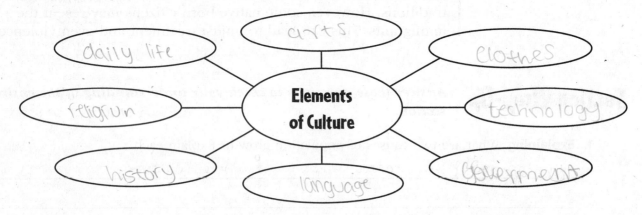

Elements of Culture: daily life, arts, clothes, religion, technology, history, language, government

Notes | Read to Learn

What Is Culture? (pages 83–86)

Listing

What social groups are included in a culture?

- young
- old
- male
- female

Culture is the way of life of a group of people who share similar beliefs and customs. Some of the shared cultural elements include language, religion, history, daily life, arts, and government.

Social Groups

Scientists study culture by looking at different groups in a society, such as the young and the old, and males and females. In all cultures, the family is the most important social group. Another type of group is an **ethnic group.** People in an ethnic group share a language, history, religion, and physical traits.

Language, Religion, History, Daily Life, Arts

Sharing a language unifies people in a culture. A language can be spoken in different ways, however. A **dialect** is a local form of a language with unique vocabulary and pronunciation. More than 2,000 languages are spoken throughout the world.

Five major religions and hundreds of other religions are practiced on Earth. Religious beliefs help people answer questions about life's meaning. A shared history also helps a culture define what is important.

What Is Culture? (continued)

Labeling

Indicate whether the types of government below are limited or unlimited.

1. democracy
   unlimited

2. dictatorship
   limited

3. monarchy
   limited & unlimited

Cultures vary in the types of food the people eat, the clothing they wear, and the types of homes they build. In addition, music, painting, and other arts reflect what a culture thinks is beautiful and meaningful.

Government and Economy

Governments create rules so people can live together without conflict. In a **democracy,** the government is limited, and the people hold power. A **dictatorship** is a type of unlimited government in which a leader or dictator rules the country, often by force. A government in which a king or queen inherits power is called a **monarchy.** Monarchs at one time had unlimited power. Today most monarchies are constitutional, or share power with elected officials.

A country's economy includes the ways people earn a living, use resources, and trade with other countries. If an economy is successful, most of the people have a good quality of life.

Cultural Change (pages 86–87)

Identifying

Identify four ways that cultures spread.

1. _resources_
2. _Internet_
3. _television_
4. _movies_

Cultures change over time. These changes can result from technological improvements or from the influence of other cultures.

Inventions and Technology

After 8000 B.C., people learned to farm and began to settle in one place. Historians call this change the Agricultural Revolution. It allowed people to create **civilizations,** or highly developed cultures. In the 1700s, countries began to change from farming societies to industrialized societies. They used machines to make goods.

More recently, computers and other inventions have changed the way people work and communicate. Medical technology has led to people living longer. These advancements have caused cultures to change.

Cultural Diffusion

Influences from other cultures also spark cultural change. Ideas, languages, and customs are spread from one group to another through **cultural diffusion.** People have been exposed to other cultures through trade, migration, and conquest. Today the Internet, television, and movies spread cultural ideas faster than ever.

Chapter 3, Section 2

Notes | Read to Learn

Regional and Global Cultures (pages 88–89)

Finding the Main Idea

What is the main idea of this subsection?

A culture region is an area made up of several countries that share cultural traits.

Geographers divide the world into physical regions and into cultural regions. A **culture region** is an area made up of several countries that share cultural traits.

Culture Regions

The countries within a culture region usually have similar languages, histories, and ethnic groups. They also tend to share the same religion and form of government. For example, Canada and the United States make up a culture region.

Global Culture

Increased communication has torn down barriers between culture regions. The result is **globalization,** or a worldwide culture. A key feature of globalization is an interdependent economy. Countries now depend upon one another for resources and markets.

Section Wrap-Up

Answer these questions to check your understanding of the entire section.

1. **Defining** What is a culture region? Give an example of one.

   Countries that usually tend to share languages, histories, ethnic groups, religion, and form of government.

2. **Determining Cause and Effect** Why is globalization occurring?

   Interdependent economies.

Write a paragraph explaining what social groups are in general, and then describe all the social groups to which you belong.

Social groups are your family and friends. I belong to my family. I am a Muslim American. I go to Al-Rahmah school.

Chapter 3, Section 3 (Pages 92–96)
Resources, Technology, and World Trade

Big Idea

Nations of the world trade resources, creating global interdependence. As you read, complete the chart below. List three examples of each type of natural resource.

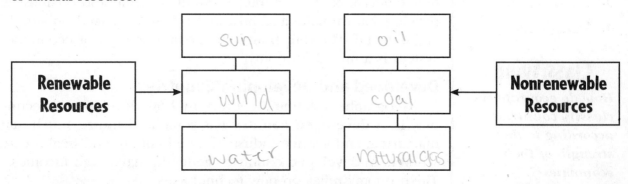

Renewable Resources: sun, wind, water
Nonrenewable Resources: oil, coal, natural gas

Notes | Read to Learn

Nativism Resurges (page 610)

Paraphrasing

Complete these sentences by filling in the blank and then circling "can" or "cannot."

1. __Renewable__ resources are available in an unlimited supply and can/(cannot) be used up.

2. __Nonrenewable__ resources are available in a limited supply and (can)/cannot be used up.

People use Earth's **natural resources** to meet their needs. Some examples of natural resources are soil, trees, wind, and oil. Natural resources provide food, shelter, goods, and energy.

There are two basic types of natural resources. **Renewable resources** cannot be used up, or they can be replaced. The sun, wind, and water cannot be used up. Forests can grow again. Rivers, the wind, and the sun can be used to produce electricity. They are valuable sources of energy.

Nonrenewable resources are limited. When these resources are used up, they cannot be replaced. Minerals such as iron ore and gold are nonrenewable resources. Fossil fuels—oil, coal, and natural gas—are also available in limited amounts, and, in fact, could run out. Fossil fuels are important sources of energy. They are used to heat homes, run cars, and generate electricity.

Chapter 3, Section 3   25

Notes — Read to Learn

Economies and Trade (pages 94–96)

Identifying

What are the four kinds of economic systems?

1. traditional
2. command
3. market
4. mixed

Classifying

How do geographers classify countries according to the strength of their economies?

1. developed country
2. developing country
3. newly industrialized country

Defining

Explain the meaning of the terms export *and* import *by using each word in a sentence.*

America exports corn and potatoes to other countries.

America imports oil from Saudi Arabia.

Economic Systems

Every society has to make economic decisions. Societies use an **economic system** to decide what goods and services to produce, how to produce them, and who will receive them.

Four kinds of economic systems exist. In a traditional economy, individuals produce goods based on traditions and customs—just like their parents and grandparents did. In a command economy, the government—not individuals—makes economic decisions. In a market economy, people and businesses decide what to make and buy, and prices are based on supply and demand. The fourth and most common type of economic system is a mixed economy.

Developed and Developing Countries

Geographers define a country by how developed its economy is. A **developed country** has some agriculture, much manufacturing, and service industries like banking and health care. Workers in developed countries generally have high incomes. The economy relies on new technologies.

Developing countries depend mainly on agriculture and have little industry. Workers generally have low incomes.

Newly industrialized countries are in the process of becoming more industrial. They are on the way to becoming developed countries.

World Trade

Resources are not evenly distributed throughout the world. Therefore, trade is important. Nations **export,** or sell to other countries, extra resources or products. Nations also **import,** or buy from other countries, resources they do not have or products they cannot make (or cannot make as cheaply).

Barriers to Trade

Trade impacts a country's economy, so governments take actions to manage international trade. Governments use trade barriers to encourage citizens to buy products made in their own country. Some governments place taxes called **tariffs** on imported goods to make those items more expensive. Another barrier to trade is a **quota,** which is a limit on the number of specific products that can be imported from a particular country.

Free Trade

Many countries are pushing for **free trade,** or the removal of tariffs and quotas. In 1992 Canada, the United States, and Mexico signed the North American Free Trade Agreement (NAFTA). This treaty removed most of the trade barriers among these countries.

Economies and Trade (continued)

Interdependence and Technology

Increased trade among countries has led to globalization. The world's people and economies have become linked together. Because of their economic **interdependence,** countries rely on each other for goods, services, and markets in which to sell their products. When economies are linked, something that happens in one location can have a global effect. For example, a drought in one country might reduce the amount of crops it can grow and sell to others, leading to shortages in countries that want to buy those crops.

Section Wrap-Up
Answer these questions to check your understanding of the entire section.

1. **Applying** Geographers define a country by how developed its economy is. Which label would be the most appropriate for the United States? Why?

   The United States is a developed country. It has lots of agriculture, manufacturing, it has a lot of money, and decent health care.

2. **Drawing Conclusions** What are two trade barriers, and why might a government want to use them?

   Tariffs and quota. Governments use them to make money.

Informative Writing
In the space provided, write a short story that explains how a war in one part of the world impacts the economy of another country far away.

In Yemen there is a war going on. Ethiopia cannot get wheat or dates. There is a shortage of their main food.

Chapter 3, Section 3

Chapter 4, Section 1 (Pages 116–122)
Physical Features

Big Idea

Geographers organize the Earth into regions that share common characteristics. In the Venn diagram below, compare landforms in the eastern, western, and interior parts of the United States and Canada.

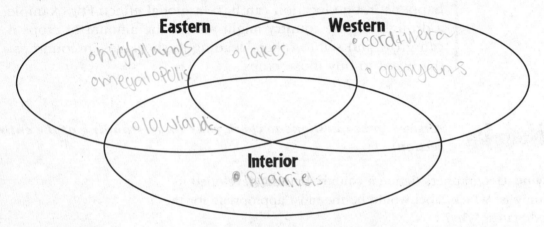

Eastern: highlands, megalopolis, lowlands
Western: cordillera, canyons
Interior: prairies
(center): lakes

Notes | Read to Learn

Major Landforms (pages 117–119)

Defining

Define the terms **contiguous** *and* **megalopolis** *by using each word in a sentence.*

The 48 states here are contiguous, or joined together inside a common boundary. Cities and suburbs along the Atlantic form a megalopolis, or connected area of urban communities.

The United States and Canada cover most of North America. The region is bordered by the Arctic Ocean in the north, the Atlantic Ocean in the east, the Gulf of Mexico in the southeast, and the Pacific Ocean in the west. Canada, the second-largest country in the world, stretches across most of the northern part of North America. The United States, the third-largest country, covers the middle part. The 48 states here are **contiguous**, or joined together inside a common boundary. Alaska and Hawaii are separate. Alaska is connected to Canada, and Hawaii lies in the Pacific Ocean about 2,400 miles from California.

A variety of landforms cover the region. Broad lowlands are found along the Atlantic and Gulf coasts. Rocky soil in the Northeast limits farming, but excellent harbors led to large shipping ports there. In the United States, cities and suburbs along the Atlantic form a **megalopolis**, or connected area of urban communities. Moving west, highland areas rise, including the Appalachian Mountains. The Appalachians run from eastern Canada to Alabama, and their rounded peaks have eroded over time. Still, rich coal deposits are mined here.

Major Landforms (continued)

Locating

What physical features are located among the cordillera?

dry basins, high plateaus, canyons, valleys

Interior lowlands are found west of the Appalachians. The horseshoe-shaped Canadian Shield wraps around Hudson Bay. Rich minerals are mined in this cold, rocky region. South of the Shield spread the Central Lowlands with grassy hills, thick forests, fertile farmland, and the Great Lakes. The Mississippi River and large cities—Chicago, Detroit, Cleveland, and Toronto, for example—are located in the Central Lowlands. West of the Mississippi stretches the Great Plains—a vast **prairie** of rolling grasslands with rich soil.

A **cordillera,** or group of mountain ranges that run side by side, towers over much of the west. The Rocky Mountains begin in Alaska and run south to New Mexico. Mountains near the Pacific coast include the Sierra Nevada, the Cascades, the Coast Ranges, and the Alaska Range. Mount McKinley—North America's highest point—rises 20,320 feet in the Alaska Range. Among the ranges lie dry basins, high plateaus, and magnificent **canyons,** or deep valleys with steep sides. The most famous is the Grand Canyon of the Colorado River.

Bodies of Water (pages 119–120)

Specifying

Where do rivers on each side of the Continental Divide flow?

West:
Pacific Ocean

East:
Arctic Ocean
Atlantic Ocean
Mississippi River

Numerous freshwater rivers and lakes are found in this region. Many are **navigable,** or wide and deep enough to allow ships to travel on them. The Great Lakes—Superior, Huron, Erie, Michigan, and Ontario—are the world's largest group of freshwater lakes. **Glaciers,** or giant ice sheets, carved them thousands of years ago. The connected lakes flow into Canada's St. Lawrence River, which empties into the Atlantic Ocean. Cities on these lakes—as well as Quebec, Montreal, and Ottawa—depend on the St. Lawrence Seaway for shipping.

The longest river in North America is the Mississippi River. It begins in Minnesota and flows 2,350 miles before emptying into the Gulf of Mexico. Products from inland port cities, such as St. Louis and Memphis, are shipped down the Mississippi to foreign ports.

Many rivers, including the Colorado and Rio Grande, flow from the Rocky Mountains. The high ridge of the Rockies forms the Continental Divide. A **divide** is a point that determines which direction rivers will flow. Rivers to the east of the Continental Divide flow toward the Arctic or Atlantic Oceans or into the Mississippi. Rivers to the west of the divide flow toward the Pacific Ocean.

Chapter 4, Section 1

Notes | Read to Learn

Natural Resources (pages 121–122)

Finding the Main Idea

As you read, write down the main idea of this subsection.

Abundant resources have enabled the U.S. and Canada to build strong industrial economies

Abundant resources have enabled the United States and Canada to build strong industrial economies. Energy resources include oil and natural gas. Texas and Alaska have large oil reserves. The United States must import oil to meet all of its needs, however. In Canada, oil and gas reserves are found near the province of Alberta. Both countries also have rich coal deposits, and hydroelectric power is generated from rivers. Niagara Falls is a major source of hydroelectricity.

The region's mineral resources are plentiful. The Rocky Mountains yield iron ore, gold, silver, and copper. Iron ore, nickel, gold, copper, and uranium are mined in the Canadian Shield.

Rich soil provides excellent farmland in parts of the region. The types of crops grown are determined by local climates, but irrigation allows even dry areas to grow many products. In California alone, farmers grow more than 200 different crops.

Forests supply timber for lumber and wood products. Large fishing industries thrive in coastal waters. The Grand Banks, once a rich fishing ground, has suffered from overfishing, however.

Section Wrap-Up

Answer these questions to check your understanding of the entire section.

1. **Explaining** In what way is the Mississippi River important to trade for the United States?

   Ships can travel on the Mississipi and some of its tributaries for great distances.

2. **Categorizing** The natural resources of the region can be divided into three broad categories. List the region's resources in their proper categories.

Energy	Mineral	Other Resources
oil gas coal Niagara Falls	iron ore uranium gold silver copper	fish trees

On a separate sheet of paper, write a paragraph describing a mountain range, a waterfall, or a canyon.

Chapter 4, Section 1

Chapter 4, Section 2 (Pages 124–128)
Climate Regions

Big Idea

The physical environment affects how people live. As you read, complete the chart below by organizing key facts about three different climate zones in the region.

Climate Zone	Location	Description
1. Tundra	northern parts of Alaska and Canada	Winters are long and cold while summers are brief and cool.
2. Pacific Coast	Southern Alaska to northern California	year round mild temperatures and abundant rainfall.
3. The East	eastern United States and Canada	humid with year round precipitation.

Notes — Read to Learn

A Varied Region (pages 125–127)

Explaining

What factors contribute to the desert climate in the inland West?

Pacific coastal mountains block humid ocean winds. Hot dry air gets trapped.

The region of the United States and Canada extends from the frozen Arctic in the far north to the steamy Tropics in the south. This results in a great variety of climates and vegetation. Most Americans and Canadians tend to live in the moderate climate zones found in the middle latitudes.

Tundra and subarctic climates are found in the northern parts of Alaska and Canada. The winters are long and cold, and the summers are short and cool. The tundra climate along the Arctic Ocean prevents the growth of trees and most plants. However, dense forests of evergreen trees grow farther south in the subarctic region.

Moist ocean winds affect the climate along the Pacific coast. A marine west coast climate is found from southern Alaska to northern California. This climate zone has mild temperatures and abundant rainfall. Southern California has a Mediterranean climate with warm, dry summers and mild, wet winters.

The inland West has a desert climate with hot summers and mild winters. Humid ocean winds are blocked from reaching this area by the mountains. In addition, hot, dry air gets trapped between the Pacific ranges and the Rocky Mountains.

A Varied Region (continued)

Applying

List the climates that support each type of plant life below.

1. Prairie grasses and grains:
humid continental

2. Variety of forests:
humid

3. Wetlands and swamps:
humid contintal

4. Rain forests:
tropical

The eastern side of the Rockies has a steppe climate. Long periods without rain, or **droughts,** provide challenges, especially to farmers and ranchers in this area. A growing population here also strains water resources.

Much of the Great Plains has a humid continental climate. Moist winds blow from the Gulf of Mexico and from the Arctic. Winters are usually cold and snowy, and summers are hot and humid. Enough rain falls to allow prairie grasses and grains to grow. The Great Plains may experience drought, however. In the 1930s, dry weather and winds turned the Great Plains into a wasteland called the Dust Bowl. The soil has since been restored through better farming methods.

The northeastern United States and parts of eastern Canada have a humid continental climate. In contrast, a humid subtropical climate is found in the southeastern United States, along with wetlands and swamps in some areas. These two climates have similar temperatures in the summer, and both have a variety of forests. In the winter, however, icy Arctic air makes the Northeast much colder.

Two parts of the United States have tropical climates. Southern Florida is in a tropical savanna zone, with hot summers and warm winters. Rain falls mostly during the summer. Hawaii is also tropical, with warm temperatures and enough rainfall to support tropical rain forests.

Natural Hazards (pages 127–128)

Identifying

Identify three weather-related natural hazards.

tornado
hurricanes
blizzards

Severe storms and other types of natural hazards pose challenges for the United States and Canada. One type of severe weather is a **tornado,** or a windstorm in the form of a funnel-shaped cloud. If the tornado touches the ground, its strong winds can knock down buildings and trees as well as lift up and move large objects. The central United States has been nicknamed "Tornado Alley" because more tornadoes form here each year than any other place in the world.

Hurricanes are wind systems that form over the ocean in tropical areas. They bring violent storms with heavy rains. Not only are the winds damaging, but hurricanes also can cause storm surges, when high levels of seawater flood low coastal areas. Hurricane season usually lasts from June to September. In August 2005, Hurricane Katrina struck the coast along the Gulf of Mexico. One of the most damaging hurricanes in history,

Natural Hazards (continued)

Determining Cause and Effect

What natural hazards are most likely to occur along the Pacific Coast? Why?

Earthquakes because fault lines of tectonic plates meet over there.

Katrina killed more than 1,800 people, wiped out hundreds of thousands of homes, and destroyed entire towns.

Severe winter storms called **blizzards** can last several hours and combine high winds with heavy snow. The snow can fall so heavily that people cannot see far, causing "white-out" conditions and making it dangerous to be outside. The snow and winds can knock down power lines and trees.

Not all natural hazards in the region are caused by the weather. Earthquakes can occur anywhere in the region, but they generally occur along the Pacific coast, where fault lines of tectonic plates meet. Volcanoes also are found where tectonic plates meet—in the Pacific ranges, southern Alaska, and Hawaii. Most of the volcanoes in this region are dormant, but several of Hawaii's volcanoes are active.

Section Wrap-Up

Answer these questions to check your understanding of the entire section.

1. **Analyzing** Where do most Americans and Canadians live? Why?

   They live in temperate climate regions.

2. **Describing** What are the characteristics of a marine west coast climate and a Mediterranean climate? Where are these two climates located in this region?

   Mediterranean is hot, dry, and like a desert. Marine west coast has hot with coastal winds. These regions are located on the west coast of the U.S.

On a separate sheet of paper, write a short story about a tornado, hurricane, blizzard, earthquake, or volcanic eruption as if you were experiencing the natural hazard.

Chapter 5, Section 1 (Pages 134–141)
History and Governments

Big Idea

The characteristics and movement of people impact physical and human systems. As you read, fill in the time line below with key events and dates in the history of the United States and Canada.

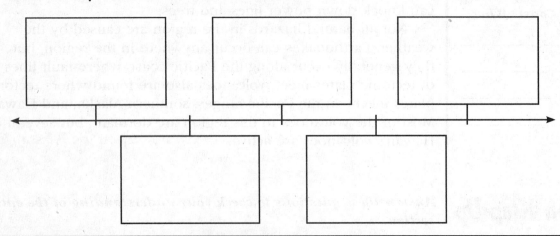

 **Read to Learn**

History of the United States (pages 135–138)

Naming

Name the three countries that established colonies in the Americas.

1. _____
2. _____
3. _____

The first people to settle the Americas were hunters from Asia who traveled with herds across a land bridge between Siberia and Alaska about 15,000 years ago. Descendants of these first settlers are called Native Americans.

Christopher Columbus reached the Americas in 1492. Soon Spain, France, and Great Britain established American **colonies,** or overseas settlements with ties to a parent country. In 1763 Great Britain gained control of France's colonies. The people in Britain's 13 coastal colonies grew resentful of British taxes and trade policies. They declared independence in 1776 and fought against British troops. Britain recognized American independence in 1783, and the United States was established.

During the 1800s, the country expanded to the Pacific Ocean, often by **annexing,** or declaring ownership of, areas of land. This expansion brought great suffering to Native Americans.

The population and economy also expanded during the 1800s. Millions of Europeans immigrated to the United States. New machines helped farmers, and the factory system produced many goods. Roads, canals, steamboats, and railroads moved goods to market quickly.

34   Chapter 5, Section 1

History of the United States (continued)

Identifying Central Issues

Over what major issue was the Civil War fought?

_____

The country became divided, however. Southern states relied on agriculture and the work of enslaved Africans. People in the north were more industrialized and criticized slavery. In 1861 several southern states withdrew from the United States. The Civil War was fought to reunite the country. Slavery ended, but racial tensions continued.

The United States became a world leader during the 1900s, fighting in World War I and World War II. After World War II, the United States and the Soviet Union struggled for world leadership in the Cold War. During this period, struggles occurred at home. Native Americans, African Americans, Latino Americans, and women sought equal rights. Since 2000, **terrorism,** or violence against civilians to reach political goals, has become a new threat. On September 11, 2001, Muslim terrorists seized four passenger planes and crashed them into New York City and Washington, D.C. About 3,000 people died.

History of Canada (pages 138–139)

Locating

Where was New France located in Canada?

_____
_____
_____
_____

Stating

What caused the Canadian colonies to unite?

_____

Canada also was first settled by Native Americans. Viking explorers from Scandinavia lived briefly on the Newfoundland coast about A.D. 1000, but they eventually left.

France and England both claimed areas of Canada in the 1500s and 1600s. The French ruled New France—the area around the St. Lawrence River and the Great Lakes. They also established the cities of Quebec and Montreal. The French became wealthy by trading with Native Americans for beaver furs, which they sold in Europe. England and France continued to fight for territory around the world in the 1600s and 1700s. By the 1760s, Britain gained control of New France.

Fearing that the United States would try to take them over, most of the colonies in Canada joined together in 1867 to become the Dominion of Canada. As a **dominion,** Canada had its own government to take care of local matters, but Britain controlled Canada's relations with other countries. The colonies became provinces, which are similar to states. Today Canada has ten provinces and three territories. The culture and language of the province of Quebec is French.

During the 1900s, Canadians fought alongside Americans and the British in the two World Wars. Canada became fully independent in 1982. The country faces the possibility that Quebec will separate and become independent.

Chapter 5, Section 1

Read to Learn

Governments of the United States and Canada (pages 140–141)

Summarizing

Complete these sentences.

In a _____ _____, voters elect their leaders.

A _____ is a type of democracy in which elected representatives choose a _____.

The United States and Canada each have a **representative democracy**—voters elect leaders who make and enforce the laws. The two systems are different in several ways, however.

The U.S. Constitution, written in the 1780s, provides the basic plan for how our national government is set up and works. Power is divided among three branches: executive, legislative, and judicial. The Constitution also set up a system of **federalism**, in which power is divided between the national and state governments.

Over the years, **amendments**, or additions, were added to the U.S. Constitution. The first 10 amendments, added in 1791, are known as the Bill of Rights. They guarantee basic freedoms.

Canada is a **parliamentary democracy.** People elect representatives to a lawmaking body called Parliament. The members of Parliament then choose a prime minister to rule.

Canada also has a federal system, with responsibilities divided between the national and provincial or territorial governments. Canada's Charter of Rights and Freedoms is similar to the U.S. Bill of Rights. It protects the liberties of Canadian citizens.

Section Wrap-Up — *Answer these questions to check your understanding of the entire section.*

1. **Explaining** What struggle occurred in the United States during the Cold War period? Who was involved in this struggle?

   _____

2. **Comparing and Contrasting** Compare and contrast the governments of the United States and Canada in the Venn diagram below.

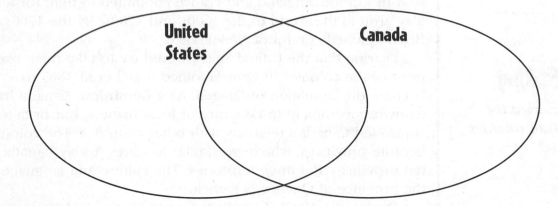

The Bill of Rights guarantees certain freedoms to the people. On a separate sheet of paper, write a paragraph explaining the freedoms that you value most.

Chapter 5, Section 2 (Pages 144–150)
Cultures and Lifestyles

Big Idea

Culture influences people's perceptions about places and regions. As you read, list key facts in the chart below about the cultures and lifestyles of the United States and Canada.

	United States	**Canada**
Language		
Art and Literature		
Leisure Time		

 **Read to Learn**

Cultures and Lifestyles of the United States (pages 145–147)

Displaying

Create a circle graph to reflect the ethnic breakdown of the U.S. population.

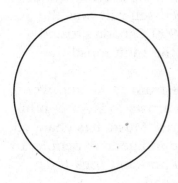

The United States has about 300 million people from different ethnic backgrounds. Early immigrants came mainly from Great Britain and Ireland. In the late 1800s, immigrants arrived from other areas of Europe, China, Japan, Mexico, and Canada. So many diverse backgrounds led some Americans to worry about cultural change. In 1882 and 1924, laws were passed that **banned,** or blocked, most immigration from China and many other countries. Immigration slowed. However, changes in U.S. laws in the 1960s increased immigration again. By 2000, nearly half of the immigrants came from Latin America and Canada, and one-third came from Asia.

People of European origin still make up two-thirds of the population. Latinos, or Hispanics, make up 15 percent and are the fasting-growing ethnic group. African Americans comprise 12 percent; Asian Americans, 4 percent; and Native Americans, 1 percent.

English is the primary language, followed by Spanish. Chinese, French, Vietnamese, Tagalog, German, and Italian are each spoken by more than 1 million people.

Most Americans follow a form of Christianity. Other religions practiced include Judaism, Islam, Buddhism, and Hinduism.

Cultures and Lifestyles of the United States (continued)

Drawing Conclusions

Why do you think Americans move to the Sunbelt?

_____
_____
_____

Artists, writers, and musicians developed distinctly American styles. Native Americans used materials from the environment to create their works, including wooden masks and pottery. Later artists focused on the beauty of the landscape or the gritty side of city life. Common themes in literature are the diversity of the people and the history and landscapes of different regions. American musicians have created many styles—folk, country, blues, jazz, rock and roll, rap, and hip-hop.

Most Americans live in cities or **suburbs,** or smaller communities around a larger city. Since the 1970s, the fastest-growing regions have been the South and Southwest—called the Sunbelt. Americans lead the world in the ownership of homes, cars, and personal computers. They also have the highest Internet usage rate. Leisure activities include watching movies and television and playing sports. Important holidays include Thanksgiving, the Fourth of July, and celebrations based on religion.

Cultures and Lifestyles of Canada (pages 149–150)

Identifying

What percentage of Canadians are of French ancestry?

_____
_____

Summarizing

Why do many people in Quebec want to be separate from Canada?

_____
_____
_____

Canada, like the United States, is made up of immigrants with many different cultures. In Canada, however, physical distances and separate cultures have led many people to feel more attached to their region than to the nation as a whole.

About 25 percent of Canadians are of French descent, and 25 percent have British origins. Another 15 percent are of other European backgrounds. Canada has more than a million **indigenous** people, or descendants of the area's first inhabitants. They are referred to as the "First Nations."

Canada is a **bilingual** country with two official languages—English and French. Many French speakers in Quebec want to become independent to better preserve their language and culture. Another cultural group that desired self-rule was the Inuit, a northern indigenous people. In 1999 Canada created the territory of Nunavut for them. There, the Inuit mostly govern themselves.

Early indigenous artists carved figures from stone and wood, made pottery, or were weavers. Canadian artists today are influenced by European and indigenous cultures. Music has changed over time, from the religious rituals of the indigenous people, to Irish and Scottish ballads in the 1700s, to pop and rock today. Movies and theater also are popular in Canada.

Foods vary by region. Seafood is common in the Atlantic Provinces. Quebec offers French cuisine. Ontario features Italian

 **Read to Learn**

Cultures and Lifestyles of Canada (continued)

Specifying

What territory was created for the Inuit?

_____

and Eastern European foods, which reflect the immigrants who settled there. Along the Pacific, British Columbia is known for salmon and Asian foods.

Canadians enjoy hockey—a sport that began in Canada. They also play lacrosse—originally a Native American game. Outdoor activities such as hunting and fishing are enjoyed as well. Canada's independence is celebrated on July 1. Like Americans, the people of Canada celebrate Thanksgiving in the fall.

Section Wrap-Up *Answer these questions to check your understanding of the entire section.*

1. **Naming** What religion do most Americans practice? What other religions are practiced in America?

_____
_____

2. **Defining** What does it mean to say that Canada is bilingual?

_____
_____

 *In the space provided, write a paragraph explaining how immigration has led to diversity in America.*

_____
_____
_____
_____
_____
_____
_____
_____

Chapter 5, Section 2

Chapter 6, Section 1 (Pages 158–162)
Living in the United States and Canada Today

Big Idea

Places reflect the relationship between humans and the physical environment. As you read, complete the diagram below. List the U.S. economic regions and provide key facts about each.

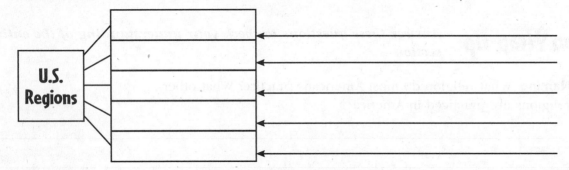

 **Notes** | **Read to Learn**

Economic Regions (pages 159–161)

Defining

Define the words profit and stock by using each one in a sentence.

_____
_____
_____
_____
_____
_____
_____

The United States and Canada have **free market** economies. People are able to buy, sell, and produce whatever they want with limited government involvement. They also can work whenever they want. Business owners produce the items they believe will have the highest **profits,** or make the most money after expenses are paid. Consumers look for the best products at the lowest prices.

Individuals can invest in businesses by buying **stock,** which represents part ownership in a company. Owning stock allows investors to share a company's profits. If the company fails, however, the stock becomes worthless. Individuals may choose to save their money in a bank. This is safer than buying stock but has less of a chance for a high financial payoff.

The United States is divided into five economic regions. Each region specializes in making products from its available resources. The Northeast region is made up of large urban areas. With little fertile soil, this region focuses on business. New York City—the country's most populous city—has many financial and media companies. Boston, Massachusetts, is a center of **biotechnology** research, or the study of cells to improve health.

Economic Regions (continued)

Listing

List six industries found in the Interior West.

1. _____
2. _____
3. _____
4. _____
5. _____
6. _____

The rich soil of the Midwest region allows farmers to grow crops such as corn, wheat, and soybeans. The region also has deposits of iron ore, coal, lead, and zinc. Cities such as Detroit and Cleveland once were centers of auto and steel manufacturing. Factories became outdated, however, and many closed. Thousands of jobs were lost.

The South has rich soil and much agriculture. Recently, the South also has experienced growing cities and industries. Textiles, electrical equipment, computers, and airplane parts are manufactured in Houston, Dallas, and Atlanta. Texas, Louisiana, and Alabama produce oil. Tourism and trade thrive in Florida.

The Interior West has long been supported by mining, ranching, and lumbering. Information technology and service industries have grown rapidly in Denver and Salt Lake City. Beautiful scenery attracts tourists to places in this region, such as Phoenix and Albuquerque.

Fruits and vegetables grow in the fertile Pacific region. Hawaii has sugarcane, pineapples, and coffee. Resources such as fish, timber, mineral deposits, and oil reserves also are plentiful. Workers in California and Washington build planes and develop software. Los Angeles is world famous for its movie industry.

Regions of Canada (pages 161–162)

Identifying

Identify the economic regions of Canada.

1. _____
2. _____
3. _____
4. _____

Canada also has distinct economic regions and a free market economy. In Canada, however, the government has a more direct role in providing services. It provides health care for citizens and regulates broadcasting, transportation, and power companies.

The Atlantic Provinces—Nova Scotia, Prince Edward Island, New Brunswick, and Newfoundland and Labrador—once had a profitable fishing industry. Overfishing, however, caused the industry to decline. As a result, many workers moved into manufacturing, mining, and tourism. Halifax, Nova Scotia, is a major shipping center.

The Central and Eastern region is made up of the provinces of Quebec and Ontario. The paper industry and hydroelectric power are two important industries in Quebec. Montreal is a major port on the St. Lawrence River, as well as a leading financial and industrial center. Foreign businesses are reluctant to invest in Quebec's economy, though, because it wants to separate from Canada.

Ontario is the wealthiest province and has the largest population. Its economic activities include agriculture, manufacturing,

Chapter 6, Section 1

Regions of Canada (continued)

Summarizing

Summarize Ontario's importance by providing five important details about it below.

1. _____
2. _____
3. _____
4. _____
5. _____

forestry, and mining. The city of Toronto is a major business and finance center. Immigrants from 170 countries have made Ontario their home.

Three provinces in the West—Manitoba, Saskatchewan, and Alberta—are known for farming and ranching. Wheat is a major export. This region also has some of the world's largest reserves of oil and natural gas. Extensive forests cover British Columbia. Lumber is used to produce **newsprint,** the type of paper used for printing newspapers. British Columbia's economy also includes mining, fishing, and tourism. Vancouver is Canada's main Pacific port.

Canada's northern region—the Yukon Territory, the Northwest Territories, and Nunavut—takes up one-third of the country. Only about 25,000 people live there, however. Many are indigenous peoples. The North has mineral deposits of gold and diamonds.

Section Wrap-Up

Answer these questions to check your understanding of the entire section.

1. **Explaining** What is a free market economy, and how does Canada's differ from that of the United States?

   _____
   _____

2. **Analyzing** In what way is the major economic activity in Canada's Atlantic Provinces changing? Why?

   _____
   _____

Imagine that you are a factory worker in the Midwest region of the United States. The factory's owners are considering shutting down the plant. On a separate sheet of paper, write a paragraph explaining why the factory should remain open.

Chapter 6, Section 2 (Pages 168–172)
Issues and Challenges

Big Idea

Cooperation and conflict among people have an effect on the Earth's surfaces. As you read, complete the outline below. Write each main heading on a line with a Roman numeral, and list important facts below it.

I. **First Main Heading** _____
   A. Key Fact 1 _____
   B. Key Fact 2 _____
II. **Second Main Heading** _____
   A. Key Fact 1 _____
   B. Key Fact 2 _____

Notes | Read to Learn

The Region and the World (pages 169–171)

Discussing

Write a brief explanation of free trade, and identify one action that the United States and Canada have taken to support it.

_____
_____
_____
_____
_____

The United States and Canada have large, productive economies. They trade with countries throughout the world. The United States, in fact, has the world's largest economy and is a leader in world trade.

The United States and Canada support free trade. They want to remove barriers in order to ease trade between countries. In 1994 the United States, Canada, and Mexico signed the North American Free Trade Agreement (NAFTA), eliminating most trade restrictions among the three countries. Today Canada is the largest trading partner of the United States, and Mexico is the second largest.

Major U.S. exports include chemicals, farm products, and manufactured goods, as well as raw materials like metals and cotton. Canada has many of the same exports, as well as seafood and timber.

Both countries also import many goods. The United States imports most of its energy resources, particularly oil. Suppliers of oil include Canada, Mexico, Venezuela, Saudi Arabia, Nigeria, and Angola.

Chapter 6, Section 2   43

Notes | Read to Learn

The Region and the World (continued)

Specifying

Write three ways the United States and Canada participate in the UN.

1. _____
_____

2. _____
_____

3. _____
_____

American consumers buy many foreign products. The United States has a huge **trade deficit**—it spends hundreds of billions of dollars more on imports than it earns from exports. This has occurred because some countries set the prices of their products low. Low prices encourage sales in the United States. At the same time, some countries put **tariffs,** or taxes, on imports to protect their own industries. Tariffs make U.S. products more expensive and reduce their sales abroad, which hurts American companies and their workers.

In contrast, Canada has a **trade surplus,** meaning that it earns more from exports than it spends on imports. Canada has a smaller population than the United States, so its energy needs are less costly.

Since the early 2000s, the United States and Canada have joined other countries to combat terrorism and violence. On September 11, 2001, terrorists attacked sites in the United States. Since then, the United States and Canada have increased border security and joined international efforts to prevent terrorist attacks.

Both the United States and Canada have important roles in the United Nations (UN), the world organization that promotes cooperation among countries in settling disputes. They provide funding, participate in agencies that provide international aid, and send soldiers to serve in UN forces.

Environmental Issues (pages 171–172)

Displaying

Complete the diagram below.

Causes
1.
2.

↓

Lower Water Levels

Effects
1.
2.
3.

The United States and Canada share environmental concerns. As people burn coal, oil, and natural gas, chemicals are released that pollute the air. Air pollution mixes with water vapor to make **acid rain,** or rain that has high amounts of chemical pollutants. Acid rain can harm trees, waterways, and the stone used in buildings. The two countries are taking steps to limit the amount of chemicals released into the air.

Another environmental concern is global warming. Some scientists think that warmer temperatures will change weather patterns, which may cause drought or melt polar ice caps. Low-lying areas like Florida may be flooded. To address this concern, Canada has passed laws to limit the amount of fossil fuels that can be burned. The United States is researching new, less harmful energy sources.

The water levels of the Great Lakes have dropped sharply because of climate changes and an increased demand for water. The lower lake levels harm fish and affect the shipping and

 **Read to Learn**

Environmental Issues (continued)

Listing

List four effects of urban sprawl.

_____

_____

_____

_____

tourism industries. The governments of both countries have urged conservation.

Another environmental challenge is **brownfields.** These abandoned places, such as factories and gas stations, contain dangerous chemicals. Until these chemicals are cleaned up, new development cannot occur at the sites. The governments of both countries have given money to communities to help clean up their brownfields.

The spread of human settlement into natural areas is called **urban sprawl.** Urban sprawl has led to the loss of farmland and wilderness, as well as increased traffic jams, pollution, and strains on water and other resources.

Section Wrap-Up *Answer these questions to check your understanding of the entire section.*

1. **Determining Cause and Effect** What actions taken by other countries have resulted in a huge U.S. trade deficit?

   _____

   _____

2. **Describing** How are the United States and Canada addressing global warming?

   _____

   _____

   _____

 *Review the information about the causes of acid rain and develop a course of action that might reduce it. On a separate sheet of paper, write a letter to a member of Congress explaining your idea and requesting that it be put into action.*

Chapter 6, Section 2

Chapter 7, Section 1 (Pages 192–196)
Physical Features

Big Idea

Geographic factors influence where people settle. As you read, complete the diagram below. Identify six key landforms in this region.

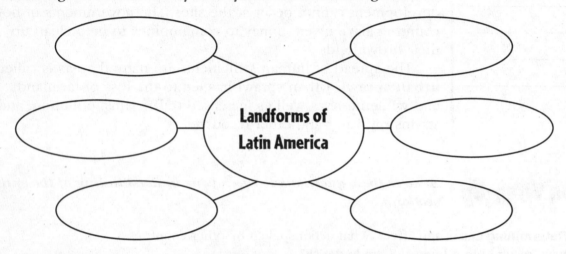

Notes — Read to Learn

Landforms (pages 193–194)

List the three groups of islands in the Caribbean and a key fact about each.

1. _____
   _____
   _____

2. _____
   _____
   _____

3. _____
   _____
   _____

Latin America is divided into three **subregions,** or smaller areas—Middle America, the Caribbean, and South America. Mexico and Central America make up Middle America. Central America is an **isthmus,** or narrow piece of land that links two larger areas of land. It links North and South America. Middle America has active volcanoes and frequent earthquakes because four tectonic plates meet there. Mountain ranges rise along Mexico's eastern and western coasts, and a high plateau lies between them. Forested mountains also form a backbone through Central America. Narrow, marshy lowlands are found along the Pacific and Caribbean coastlines.

The Caribbean islands, also called the West Indies, are divided into three groups. The Greater Antilles includes the largest islands—Cuba, Hispaniola, Puerto Rico, and Jamaica. The Lesser Antilles is an **archipelago,** or chain of islands, curving from the Virgin Islands to Trinidad. The third group, the Bahamas, is another archipelago. Some Caribbean islands are low-lying. Others, formed by volcanoes, have rugged mountains and fertile volcanic soil.

 **Read to Learn**

Landforms (continued)

The 5,500-mile-long Andes ranges and the huge Amazon Basin are major landforms in South America. The Brazilian Highlands that border the Amazon Basin end in an **escarpment,** or series of steep cliffs, that drop to the Atlantic coast. Tropical grasslands called the **Llanos** cover eastern Colombia and Venezuela. Another fertile plain called the **Pampas** stretches through much of Argentina and Uruguay.

Waterways (pages 194–195)

Explaining

Why is the Amazon River important to Latin America?

_____
_____
_____
_____

The longest river in Latin America is the Amazon. It begins in the Andes and flows about 4,000 miles to the Atlantic Ocean. Many **tributaries,** or smaller rivers, flow into the Amazon. The Amazon is used for shipping, and people also rely on it for fish.

The Paraná, Paraguay, and Uruguay Rivers form the second-largest river system in Latin America. These three rivers wind through inland areas and then flow into an **estuary,** the place where river currents meet ocean tides. This estuary—called the Río de la Plata, or "River of Silver"—flows into the Atlantic Ocean. The Orinoco River flows north through Venezuela into the Caribbean Sea.

The largest lake in South America is Venezuela's Lake Maracaibo. Lake Titicaca is located 12,500 feet above sea level high in the Andes between Bolivia and Peru. It is the world's highest navigable lake. Another key waterway in the region is the Panama Canal. This human-made waterway across the Isthmus of Panama provides a shorter route for ships traveling between the Atlantic and Pacific Oceans.

A Wealth of Natural Resources (pages 195–196)

Discussing

What is gasohol, and why does Brazil produce it?

_____
_____
_____

Brazil is the largest country in Latin America and has the most natural resources. Rain forests cover more than half of the country and provide timber, rubber, palm oil, and Brazil nuts. Mineral deposits in Brazil include bauxite, gold, tin, iron ore, and manganese. Brazil does not have large oil reserves. To limit dependence on imported oil, Brazil produces a fuel for cars called **gasohol,** an alcohol made from sugarcane and gasoline.

Venezuela and Mexico produce enough oil and natural gas to meet their needs and to export to other countries. Bolivia and Ecuador also have oil and natural gas deposits. Other minerals

Chapter 7, Section 1

A Wealth of Natural Resources (continued)

Identifying

Underline the reasons Nicaragua and Guatemala have difficulty mining gold.

in the region include silver in Mexico and Peru. Venezuela has rich iron ore deposits. Colombia mines emeralds, and Chile exports copper.

The Caribbean islands have few mineral resources. Jamaica is an exception. It has large deposits of bauxite, which is used to make aluminum. Cuba mines nickel, and the Dominican Republic mines gold and silver. Nicaragua and Guatemala in Central America have rich gold deposits. However, political conflicts and transportation difficulties make mining their gold difficult.

Section Wrap-Up

Answer these questions to check your understanding of the entire section.

1. **Explaining** What is the significance of the Panama Canal?

   _____

   _____

2. **Organizing** Complete this chart by listing the country(ies) in which the mineral resources are found.

Minerals	Country(ies)
Bauxite	
Copper	
Emeralds	
Gold	
Iron ore	
Manganese	
Nickel	
Silver	
Tin	

On a separate sheet of paper, write a paragraph describing some of the challenges of living in an area with active volcanoes and frequent earthquakes.

Chapter 7, Section 2 (Pages 198–202)
Climate Regions

Big Idea

The physical environment affects how people live. As you read, use the Venn diagram below to compare and contrast the tropical rain forest and the tropical savanna climate zones.

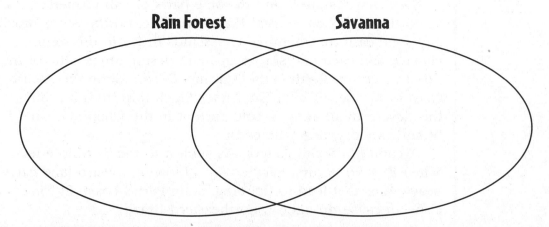

 **Notes** | **Read to Learn**

Hot to Mild Climates *(pages 199–201)*

Paraphrasing

As you read, complete these sentences.

The Tropics are generally warm year-round because the area receives _____.

Two factors that affect the climate in Latin America are

_____

and _____.

Most of Latin America is located in the area between the Tropic of Cancer and the Tropic of Capricorn. This area is called the **Tropics.** It receives direct sunlight for much of the year, and the temperatures are generally warm. However, mountain ranges and wind patterns contribute to a variety of climates in the region.

Some Caribbean islands and much of Central America and South America have a tropical wet climate. Temperatures are hot and rainfall is heavy throughout the year. Much of this climate zone is covered by **rain forests,** or dense stands of trees and other plants that thrive on high amounts of rain.

The world's largest rain forest is in South America's Amazon Basin. The trees there grow so close together that their tops form a **canopy,** or an umbrella-like covering of leaves. The canopy blocks most sunlight from reaching the forest floor.

Most Caribbean islands, parts of Middle America, and north central South America have a tropical dry, or savanna, climate zone. Temperatures are hot, and rainfall is abundant, but this climate zone also has a long dry season.

Hurricanes often strike the Caribbean islands from June to November. These storms can cause much damage.

Chapter 7, Section 2   49

Hot to Mild Climates (continued)

Identifying

Identify two effects of El Niño.

1. _____
_____
_____

2. _____
_____
_____

The areas south of the Tropic of Capricorn have temperate climates. A humid subtropical climate—with short, mild winters and long, hot, humid summers—covers southern Brazil and the Pampas of Argentina and Uruguay. Central Chile has a Mediterranean climate, with dry summers and rainy winters. Farther south is a marine coastal climate, with heavier rainfall throughout the year.

Dry climates are found in some parts of Latin America, such as northern Mexico, coastal Peru and Chile, northeastern Brazil, and southeastern Argentina. Grasslands thrive in the steppe climate, and cacti and shrubs grow in desert zones. One of the driest places on Earth is the Atacama Desert along the Pacific coast of northern Chile. The Andes block rain from reaching this desert. In addition, a cold current in the Pacific Ocean brings only dry air to the coast.

Weather in South America is subject to the El Niño effect. When El Niño occurs, Pacific winds blowing toward land carry heavy rains that lead to flooding along Peru's coast. El Niño also can cause drought in northeastern Brazil.

Elevation and Climate (pages 201–202)

Labeling

Label the four altitude zones in the Andes.

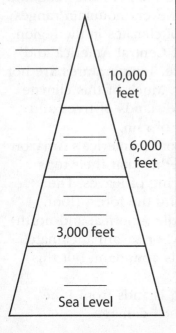

A place's height above sea level, called **altitude**, affects its climate. Higher altitudes have cooler temperatures, even in warm regions such as the Tropics. In South America, the Andes have four altitude zones of climate.

The *tierra caliente,* or "hot land," is named for the hot and humid areas that are near sea level. Farmers here grow bananas, sugarcane, and rice. The *tierra templada,* or "temperate land," is found from 3,000 feet to 6,000 feet. It is moist and pleasant, with mild temperatures. *Tierra templada* is the most densely populated climate zone. People grow corn, wheat, and coffee in this zone.

The *tierra fría,* or "cold land," extends from 6,000 feet to 10,000 feet. Average temperatures in this zone can be as low as 55°F. Crops that do well in these conditions include potatoes, barley, and wheat. The highest altitude, starting at about 10,000 feet, is called *tierra helada,* or "frozen land." Conditions are harsh, vegetation is sparse, and few people live in this high altitude.

Section Wrap-Up

Answer these questions to check your understanding of the entire section.

1. **Organizing** Complete this chart by identifying the two tropical climate zones and five additional climate zones found in Latin America.

Tropical Climate Zones	Other Climate Zones

2. **Explaining** What factors contribute to the Atacama Desert being so dry?

_____

_____

In the space provided, write a paragraph explaining why most people live in the tierra templada climate zone rather than in the other three altitude zones in the Andes.

_____
_____
_____
_____
_____
_____
_____
_____
_____
_____
_____
_____
_____
_____

Chapter 8, Section 1 (Pages 208–215)
History and Governments

Big Idea

All living things are dependent upon one another and their surroundings for survival. As you read, complete the chart below. List key facts about the civilizations of the region.

	Key Facts
Olmec	
Maya	
Toltec	
Aztec	
Inca	

Notes — Read to Learn

Early History (pages 209–211)

Listing

List four achievements of the Maya.

1. _____
2. _____
3. _____
4. _____

Stating

How many people lived in Tenochtitlán?

_____

The Olmec built Latin America's first civilization in southern Mexico. It lasted from 1500 B.C. to 300 B.C. Some Olmec cities grew **maize,** or corn, as well as squash and beans. Some cities controlled mineral resources such as **jade,** a green semiprecious stone, and **obsidian,** a hard, black, volcanic glass used to make weapons. Other cities were religious centers.

The Maya lived in Mexico's Yucatán Peninsula from about A.D. 300 to A.D. 900. They built huge pyramid temples, used astronomy to develop a calendar, and had a number system based on 20. They also used **hieroglyphics,** a form of writing that uses signs and symbols, to record their history. Around A.D. 900, the Maya civilization mysteriously collapsed. Around the same time, the Toltec conquered northern Mexico. They controlled trade and held a monopoly on obsidian, giving them the most powerful weapons in the region.

The Aztec arrived around 1200. They adopted Toltec culture, conquered neighboring peoples, and took control of trade. Their capital, Tenochtitlán, was built on an island in a lake. About 250,000 people lived there. Roads and bridges linked the city to the mainland.

52 Chapter 8, Section 1

Early History (continued)

Determining Cause and Effect

How did Hernán Cortés defeat the Aztec?

_____
_____
_____
_____

The Inca were powerful in Peru during the 1400s. Their **empire,** or large territory with many different peoples under one ruler, stretched 2,500 miles along the Andes. Roads and suspension bridges linked all parts of the empire to Cuzco, the capital. The Inca also had military posts, irrigation systems, and a complex system of record keeping.

Spanish explorers reached Latin America in the late 1400s. In 1521 Hernán Cortés defeated the Aztec and their simple weapons with guns, cannons, and horses. Diseases also wiped out the Aztec and Inca. In 1532 Francisco Pizarro attacked and quickly conquered the Inca Empire. These conquests brought Spain and its new empire enormous wealth. Other European countries also seized different parts of the Americas and established colonies. Portugal took over Brazil. The French, British, and Dutch settled Caribbean areas.

The Europeans set up colonial governments, spread Christianity, and forced Native Americans to grow **cash crops,** or farm products grown for export. After many Native Americans died from disease, enslaved Africans were brought to work on plantations and in mines.

Forming New Nations (pages 212–215)

Summarizing

What conflicts arose after independence?

1. _____
2. _____
3. _____

Naming

Name three exports of Latin America in the late 1800s.

1. _____
2. _____
3. _____

Inspired by the American and French revolutions, the people of Latin America fought for their freedom. In Haiti, Toussant-Louverture led a revolt that overthrew French rule in 1804. Simón Bolívar fought against Spain and won freedom for Venezuela, Colombia, Ecuador, and Bolivia in 1819. Mexico gained independence in 1821. José de San Martín helped lead Chile and Peru to freedom. Brazil broke away peacefully from Portugal in the 1820s.

After winning independence, Latin America faced political and economic challenges. Slavery was ended, but conflicts occurred over the role of religion, boundary lines, and the huge gap between rich and poor. Strong leaders known as **caudillos** were supported by the upper class and often made it difficult for democracy and prosperity to grow.

In the late 1800s, Latin America's economy depended on agriculture and mining. Foreign companies moved in to control the export of bananas, sugar, coffee, copper, and oil. The United States also increased its political influence in the region. It fought Spain to gain Cuba in 1898. Puerto Rico came under U.S. control. The United States helped Panama win its independence from

Chapter 8, Section 1

 **Read to Learn**

Forming New Nations (continued)

Determining Cause and Effect

What resulted from Latin America's increasing debt?

_____
_____
_____
_____

Colombia in 1903. In return, Panama allowed America to build the Panama Canal. U.S. troops entered Haiti, Nicaragua, and the Dominican Republic to protect American economic interests. Many Latin Americans feared that the United States would try to control them. In response, the United States established the Good Neighbor Policy in the 1930s, promising not to send military forces and to respect Latin American rights.

In the mid-1900s, Latin American leaders borrowed money from other countries to encourage industrial growth. The increasing debt weakened local economies, however. Many people lost jobs and faced rising prices. Dissatisfied groups in some countries rebelled against ruthless leaders. In Cuba, Fidel Castro led a successful revolt and set up a **communist state,** in which the government controls the economy and society. Civil wars raged in other countries, such as El Salvador.

Economic and political reforms in the 1980s strengthened many Latin American countries. However, challenges still exist in the region. These include rapid population growth, limited resources, the illegal drug trade, and the vast division between the wealthy and the poor. Leaders elected in the early 2000s promised significant changes.

Section Wrap-Up

Answer these questions to check your understanding of the entire section.

1. **Explaining** What is obsidian? How did it help make the Toltec powerful?
   _____
   _____

2. **Sequencing** Complete the time line below with key events and dates in the history of Latin America. Extend the line and add more boxes if necessary.

 *On a separate sheet of paper, write a paragraph explaining whether you think U.S. involvement in Latin America has helped or hurt the region.*

Chapter 8, Section 2 (Pages 218–224)
Cultures and Lifestyles

Big Idea

The characteristics and movement of people impact physical and human systems. As you read, complete the diagram below. Add one or more facts to each of the outer boxes.

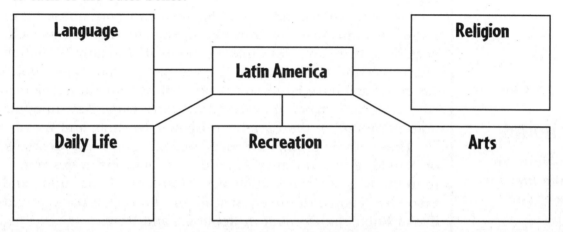

 Read to Learn

The People (pages 219–221)

Summarizing

Summarize three types of migration that occur in Latin America.

1. _____

2. _____

3. _____

Latin America has a high population growth rate. Central America has the fastest-growing populations. Guatemala and Honduras are expected to double their populations by the year 2050.

The region's climates and landscapes affect where people live. Areas with extreme temperatures, rain forests, deserts, and mountains are sparsely populated. Most people live in moderate climates along the coasts of South America or in an area stretching from Mexico into Central America. These areas have fertile soil and allow for easy movement of people and goods.

Latin America's population has been affected by **migration,** or the movement of people. People immigrate into the region in search of jobs or personal freedom. At the same time, some Latin Americans leave the region to escape political unrest or to find a better way of life. Many legally and illegally enter the United States looking for work. Other people move within the region, leaving their farms to search for jobs in rapidly growing cities.

Some of the world's largest cities are in Latin America, including Mexico City, São Paulo, Rio de Janeiro, and Buenos Aires. Millions of villagers that stream to these cities face poverty, crowded neighborhoods, lack of sanitation, and rising crime.

Chapter 8, Section 2

The People (continued)

Locating

Where do most Mestizos live in this region?

_____
_____
_____

Ethnic groups in Latin America include Native Americans, Europeans, Africans, Asians, and people of mixed descent. Mexico, Central America, Ecuador, Peru, and Bolivia are home to most of the region's Native Americans. They try to maintain their languages and traditions while adopting features of other cultures.

Since the 1400s, millions of Spanish and Portuguese have settled in Latin America. Other Europeans immigrated as well. Argentina, Uruguay, and Chile are populated mainly by people of Spanish and Italian origin. The Caribbean islands and northeastern Brazil have large populations of African Latin Americans who are descendants of enslaved Africans. Large Asian populations are found in the Caribbean islands, Guyana, and Brazil.

Defining

Define pidgin language and give an example of one.

_____
_____
_____
_____

These ethnic groups blended over the centuries. **Mestizos**, or people of mixed Native American and European descent, form the largest groups in Mexico, Honduras, El Salvador, and Colombia. People of mixed African and European backgrounds live in Cuba, the Dominican Republic, and Brazil.

Spanish is the most widely spoken language, although most Brazilians speak Portuguese. Quechua, spoken centuries ago by the Inca, is an official language of Peru and Bolivia. English and French are spoken on some Caribbean islands. Several countries developed a **pidgin language** by combining parts of different languages. Haiti's Creole, for example, is a mix of French and African languages.

Daily Life (pages 223–224)

Listing

What sports are popular in Latin America?

_____
_____
_____

Christianity plays a significant role in Latin American cultures. Most people became Christians during colonial times. Other faiths include traditional Native American and African religions, Islam, Hinduism, Buddhism, and Judaism.

Family is central to the Latin American way of life. Multiple generations often live in the same house, and extended families tend to live near each other. The father is the leader and decision maker, although the mother is the leader in some parts of the Caribbean.

Sports are widely popular. Soccer is the primary sport, and baseball also has a strong following. Cuba was the second country to play baseball, after the United States. Many skilled baseball players have entered the U.S. professional leagues. Cricket is played in Caribbean countries that were once ruled by the British.

Notes | Read to Learn

Daily Life (continued)

Discussing

List two holidays celebrated in Latin America and a key fact about each.

1. _____
_____

2. _____
_____

Religious and patriotic holidays are common. Many countries hold a large festival called **carnival** on the day before Lent begins. In Mexico, a holiday called Day of the Dead is celebrated in honor of family members who have died.

Foods blend the traditions of the region. Corn and beans are common in Mexico and Central America. Beans and rice make up the main diet of Caribbean islanders and Brazilians. Beef is the national dish in Argentina, Uruguay, and Chile.

Music and art also reflect the region's ethnic mix. For example, Cuban music uses African rhythms. In the 1930s, Mexican artists painted **murals,** or large paintings on walls, that were similar to the artistic traditions of the Maya and Aztec. Magic realism is a writing style invented by Latin American writers of the late 1900s. It combines fantastic events with the ordinary.

Section Wrap-Up

Answer these questions to check your understanding of the entire section.

1. **Explaining** Why are Latin American cities growing so rapidly?
   _____
   _____

2. **Organizing** Complete this chart to show the regional distribution of ethnic groups.

Origins	Country(ies)
African descent	
Asian descent	
European descent	
Native American	

On a separate sheet of paper, write a paragraph describing a holiday you invent that can be celebrated with your family and friends. Identify the reasons for the holiday and why it is important to invite others.

Chapter 8, Section 2

Chapter 9, Section 1 (Pages 232–236)
Mexico

Big Idea

Patterns of economic activities result in global interdependence. As you read, complete the chart below with key facts about Mexico's economic regions.

Region	Key Facts
North	
Central	

Notes | Read to Learn

Mexico's People, Government, and Culture (pages 233–234)

Summarizing

List three facts about Mexico's government.

1. _____
2. _____
3. _____

The population of Mexico is a mix of Spanish and Native American heritage. About two-thirds of the people are mestizos, and one-fourth are Native American. Rural traditions are strong, but 75 percent of the people live in cities. The capital and largest city, Mexico City, has nearly 22 million people. Reflecting Spanish culture, Mexico's cities are organized around public squares called **plazas**. These serve as centers of public life.

Mexico is a federal republic with power shared between the national and state governments. A strong national president serves only one six-year term. He or she has more power than the legislative and judicial branches. A revolution occurred in Mexico in the early 1900s. After that, one political party ruled the country for a long time. In the 1990s, people were frustrated by economic troubles and their lack of political power. In 2000 Mexicans elected a president from a different political party for the first time in 70 years.

Mexican culture has been influenced by Native Americans and Europeans. Folk arts such as wood carving reflect Native American traditions. European culture is seen in sports such as soccer. Carved and painted religious statues blend the cultures.

Mexico's People, Government, and Culture (continued)

Stating

What is a popular Mexican food?

_____

Artists and writers are national treasures. Diego Rivera and his wife, Frida Kahlo, were famous painters of the early 1900s. Famous authors include Carlos Fuentes and Octavio Paz, who wrote about the values of Mexico's people.

Mexicans hold celebrations called fiestas, which are highlighted with parades, fireworks, music, and dancing. Popular foods—in both Mexico and the United States—include tacos and enchiladas.

Mexico's Economy and Society (pages 234–236)

Comparing

What crops are grown in the North and South regions of Mexico?

North:

_____
_____
_____
_____

South:

_____
_____
_____
_____

Identifying

Which region is the most heavily populated?

_____

Mexico has a growing economy. Three distinct economic regions result from the country's physical geography and climate. The North has dry and rocky land. Farmers must use irrigation to grow cotton, grains, fruits, and vegetables for export. Grasslands support cattle ranches worked by cowhands called **vaqueros.** The North also has deposits of copper, zinc, iron, lead, and silver. Factories are located near the Mexico–United States border in cities such as Monterrey, Tijuana, and Cuidad Juárez. Many factories are **maquiladoras,** or foreign-owned plants that hire Mexican workers to assemble parts made in other countries. The finished products are then exported.

More than half of Mexico's people live in the Central region. It has a pleasant climate and fertile volcanic soil, allowing for productive farming. Workers in industrial cities such as Mexico City and Guadalajara make cars, clothing, household items, and electronic goods. The country's energy industry is centered along the Gulf of Mexico, near offshore oil and gas deposits.

Mexico's poorest economic region is the South. **Subsistence farms,** or small plots where farmers grow only enough food to feed their families, are common in the mountains. On coastal lowlands, wealthy farmers grow sugarcane and bananas on **plantations,** or large farms that raise a single cash crop. Coastal resorts such as Acapulco and Cancun attract tourists.

Mexico's economy is shifting in priority from agriculture to manufacturing. The North American Free Trade Agreement (NAFTA) helped Mexico increase trade with Canada and the United States. Factories there produce steel, cars, and consumer goods. Banking and tourism industries also contribute to the economy. Economic advances have raised the standard of living, especially in the North.

Chapter 9, Section 1

Notes | Read to Learn

Mexico's Economy and Society (continued)

Categorizing

Write a positive outcome and a negative outcome of Mexico's economic growth.

_____
_____
_____
_____

Economic growth also has raised concerns about the environment and dangers to the health and safety of workers. Pollution has increased, and smog often blankets Mexico City.

Mexico's population is growing rapidly. As people move to cities in search of jobs, many have crowded together in slums. Some Mexicans are **migrant workers** who travel to find work planting or harvesting crops. Migrant workers often cross the border into the United States, sometimes illegally, to work.

Many Native Americans are poor and live in rural areas. In the 1990s, Native Americans in southern Mexico rose up against the government and demanded changes to improve their lives. Their struggle has not been resolved.

Section Wrap-Up

Answer these questions to check your understanding of the entire section.

1. **Identifying** Provide three examples of Native American and European influences on Mexican culture.

   _____
   _____

2. **Describing** How has Mexico's economy changed in recent years?

   _____
   _____

Persuasive Writing

Choose one of the challenges facing Mexico. In the space provided, write a newspaper editorial in which you suggest steps Mexico's government could take to improve the situation.

_____
_____
_____
_____
_____

Chapter 9, Section 2 (Pages 237-240)
Central America and the Caribbean

Big Idea

The physical environment affects how people live. As you read, compare and contrast Guatemala and Costa Rica in the Venn diagram below.

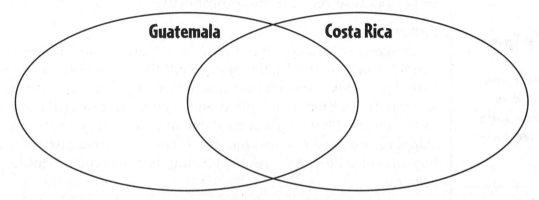

Notes | Read to Learn

Countries of Central America (pages 238-239)

Specifying

Write down two economic changes in Guatemala in recent years.

1. _____
   _____
2. _____
   _____

Belize, Guatemala, El Salvador, Honduras, Nicaragua, Costa Rica, and Panama make up Central America. Most people in these countries farm. Bananas, sugarcane, and coffee are exported. Political and ethnic conflicts have weakened some economies in this region.

Guatemala

Half the people in Guatemala are descended from the ancient Maya. Many are of mixed Maya and Spanish heritage. Both languages are spoken in the country. A small but wealthy and powerful group owns most of the land in Guatemala. In the late 1990s, conflict erupted as rebel groups fought for control of the land. More than 200,000 people were killed or missing after the conflict ended.

Guatemala's economy has recently undergone some changes. Farmers are beginning to grow more valuable cash crops, such as fruits, flowers, and spices. In the early 2000s, Guatemala joined other Central American countries in a free trade agreement with the United States. This should allow the people to sell more goods to the United States.

Countries of Central America (continued)

Differentiating

How does Costa Rica differ from its neighbors?

_____

_____

Costa Rica

Costa Rica has a stable democratic government. It has fought no wars since the 1800s. A police force—but no army—keeps law and order in the country. Costa Rica has fewer poor people than other countries in Central America. Costa Ricans have a high **literacy rate,** which is the percentage of people who can read and write. Literate workers generally earn higher wages because they are more productive.

Panama

Panama is located on the narrowest part of the isthmus of Central America. The United States built the Panama Canal there, which provides a shorter and faster route between the Atlantic and Pacific Oceans. The United States gave control of the canal to Panama in 1999. Panama earns money from the fees it charges shipping companies to use the canal. The canal area also attracts buyers and sellers. As a result, Panama has become a banking center.

Calculating

The Panama Canal was completed in 1914. For how long did the United States control it?

_____

Countries of the Caribbean (pages 239–240)

Stating

What happens to Cubans who criticize the government?

_____

_____

Some Caribbean countries, such as Cuba and Haiti, face political and economic challenges. Others, like Puerto Rico, are more stable.

Cuba

Cuba lies 90 miles south of Florida. It has a **command economy**—the communist government determines how resources are used and what goods and services are produced. The economy has not been successful, however, and most Cubans are poor. For many years, Cuba's economy relied on the sale of a single crop—sugar. The government is now trying to develop tourism and other industries to end the dependence on sugar.

Cuba's longtime dictator, Fidel Castro, controls society. People who criticize the government are often jailed. The United States condemns Cuba for these actions.

Haiti

Haiti is located on the western side of the island of Hispaniola. It has a troubled history. The government is unstable because of ongoing conflicts between political groups. Most Haitians live in poverty. A vital source of income is **remittances,** or money sent back home by Haitians who work in other countries.

Notes | Read to Learn

Countries of the Caribbean (continued)

Summarizing

Summarize Puerto Rico's economy.

Products made:

_____

_____

_____

Crops grown:

_____

_____

Other:

_____

Puerto Rico

Since 1952, Puerto Rico has been a **commonwealth**, or a self-governing territory of the United States. The people are American citizens who can travel freely between Puerto Rico and the United States.

Puerto Rico has a higher standard of living than other countries in the Caribbean. Factories produce medicines, machinery, and clothing. Farmers grow sugarcane and coffee. The tourism industry also thrives in Puerto Rico.

Section Wrap-Up *Answer these questions to check your understanding of the entire section.*

1. **Analyzing** What is the significance of the high literacy rate in Costa Rica?

_____

_____

2. **Explaining** Why is the Panama Canal important to the economy of Panama?

_____

_____

Expository Writing *In the space provided, write a paragraph explaining how Cuba's communist government has affected the Cuban economy and society.*

_____

_____

_____

Chapter 9, Section 2

Chapter 9, Section 3 (Pages 246–252)
South America

Big Idea

People's actions can change the physical environment. As you read, describe Brazil's economy on the diagram below. Write the main idea on the single line and supporting details on the lines to the right.

Notes | Read to Learn

Brazil (pages 247–249)

Defining

What are favelas, and why have they emerged in Brazil?

_____

_____

_____

Identifying

What does Brazil use to make a substitute for gasoline?

_____

Brazil is the largest country in South America. Its culture is largely Portuguese rather than Spanish. Brazil's 187 million people are of European, African, Native American, Asian, and mixed ancestry. Most live in cities along the Atlantic coast, such as São Paulo and Rio de Janeiro. Millions have moved to coastal cities in search of jobs, settling in overcrowded slum areas called **favelas.** To reduce crowding, the government is encouraging people to move back to less-populated areas. The capital, Brasília, is located 600 miles inland.

Brazil is the world's leading producer of coffee, oranges, and cassava. Agricultural output has grown as more land has been cleared to grow crops. Machinery is used to perform many tasks, and crops have been scientifically changed to produce more and prevent disease.

Valuable minerals mined in Brazil include iron ore, bauxite, tin, manganese, gold, silver, and diamonds. Energy resources include offshore oil deposits and hydroelectric power. Sugarcane is used to make a substitute for gasoline. Booming industries produce machinery, airplanes, cars, food products, medicines, paper, and clothing.

Brazil (continued)

Listing

What economic activities occur in the selva?

_____

_____

_____

Brazil's greatest natural resource is the Amazon rain forest, called the **selva**. To promote economic development, the government has encouraged mining, logging, and farming in the rain forest. However, deforestation harms the rain forest's ecosystem and biodiversity, reduces the amount of oxygen released, and may affect Earth's climate patterns. Brazil has agreed to protect some rain forest areas.

The country is a democratic federal republic. Citizens elect the president and other leaders. Brazil's national government is much stronger than its 26 state governments.

Argentina (pages 249–250)

Paraphrasing

Complete these sentences.

Argentina had a high

_____

because it borrowed money from

_____.

Argentina was unable to make all of its

_____,

so it had to

_____

on its debts.

Argentina is the second-largest country in South America. The Andes tower in the west. Central Argentina has vast grasslands known as the Pampas. Most people have Spanish and Italian origins, and more than one-third live in the beautiful capital, Buenos Aires. This bustling port has been nicknamed "the Paris of the South."

Farming and ranching are vital to Argentina's economy. Cowhands called **gauchos** tend livestock on the Pampas. Gauchos, the national symbol of Argentina, are admired for their independence and horse-riding skills. Beef and beef products are the country's main exports. Argentina's factories produce food products, cars, chemicals, and textiles. Zinc, iron ore, copper, and oil are mined in the Andes.

In the 1990s, Argentina borrowed money from foreign banks, leading to a high **national debt**, or money owed by the government. Argentina had to **default** on its debts, meaning that it missed debt payments to the banks that lent the money. The economy has since recovered, and part of the debt has been repaid.

After gaining independence in the early 1800s, Argentina was led by military leaders. Today Argentina is a democratic federal republic. A powerful president is elected every four years.

Other Countries of South America (pages 251–252)

Venezuela, located along the Caribbean Sea, is a leading producer of oil and natural gas. Factories make steel, chemicals, and food items. Farmers grow sugarcane and bananas or raise cattle. Yet many Venezuelans are poor. Some live in slums

Chapter 9, Section 3

Notes | Read to Learn

Other Countries of South America (continued)

Specifying

What was supposed to improve the lives of poor Venezuelans?

_____

Stating

What are the key elements of Venezuela and Chile's economies?

Venezuela:

_____
_____

Chile:

_____
_____

surrounding Caracas, the capital. In 1998 Venezuelans elected Hugo Chávez as president. A former military leader, he promised to use oil money to improve the lives of the poor. However, his strong rule has divided the country.

Colombia borders both the Caribbean Sea and the Pacific Ocean. Most people live in the valleys and plateaus of the Andes. Bogotá, the capital, is located on an Andean plateau. The country mines coal, oil, and copper and is the world's leading supplier of emeralds. It exports bananas, sugarcane, rice, cotton, and world-famous coffee. Despite its economic strengths, Colombia faces much unrest. Wealth is in the hands of a few, while many people are poor. Drug dealers pose another problem. They pay farmers to grow coca leaves to produce cocaine.

Long, ribbon-shaped Chile borders the Pacific Ocean. Its landscapes vary from dry deserts in the north, to central fertile valleys, to glaciers in the south. Chile's economy is based on mining copper, gold, silver, iron ore, and **sodium nitrate,** a mineral used in fertilizer and explosives. In addition, farmers raise wheat, corn, beans, sugarcane, potatoes, grapes, and apples. People also raise cattle and sheep or work in the large fishing industry. Like Argentina, Chile was under military rule for many years. It is now a democracy.

Section Wrap-Up

Answer these questions to check your understanding of the entire section.

1. **Explaining** Why has Brazil's agricultural output increased?

_____
_____

2. **Comparing** Complete this chart by listing the types of governments.

Country	Type of Government
Brazil	
Argentina	
Chile	

Expository Writing

On a separate sheet of paper, write a paragraph explaining why people in other areas of the world are concerned when Brazilians harm their rain forest.

Chapter 10, Section 1 (Pages 274–280)
Physical Features

Big Idea

Geographic factors influence where people settle. As you read, complete the chart below with key facts about the landforms, waterways, and resources of Europe.

Features	Key Facts
Landforms	
Waterways	
Resources	

Notes — Read to Learn

Landforms and Waterways (pages 275–277)

Listing

What are five types of landforms found in Europe?

1. _____
2. _____
3. _____
4. _____
5. _____

The continent of Europe shares a common landmass with Asia. This landmass is called Eurasia. Europe is located on the western portion of Eurasia.

Europe has a long coastline. It borders the Atlantic Ocean and the Baltic, North, Mediterranean, and Black Seas. Only a few countries are **landlocked,** or do not border an ocean or a sea. Nearness to water has influenced Europe's history and people. Shipping and fishing encouraged trade and helped build Europe's economy. Exploration spread European culture worldwide and brought ideas from Asia, Africa, and the Americas to Europe.

Peninsulas and Islands

Europe is a huge peninsula. It has many smaller peninsulas branching out from it. Europe also has many islands, including Great Britain, Ireland, Iceland, and Cyprus. At one time, seas, rivers, and mountains separated people living on these peninsulas and islands. Thus, many different cultures developed.

Landforms and Waterways (continued)

Defining

What does the word navigable *mean?*

_____
_____
_____
_____
_____

Plains

The Northern European Plain is Europe's major landform. The soil is rich, and the plain also holds underground deposits of coal, iron ore, and other minerals. Most of Europe's population live and work on this vast plain. Other lowland areas include the Hungarian Plain and Ukrainian Steppe.

Mountains and Highlands

Europe's highest mountain ranges form the Alpine Mountain System. The Alps, the Pyrenees, and the Carpathians are included in this system. People and goods travel through **passes,** or low areas between mountains. Several other highland areas are used for mining and grazing livestock.

Waterways

Many rivers, lakes, and other waterways are found in Europe. The Danube and Rhine are two of Europe's longest rivers. Many of the rivers are **navigable**—wide and deep enough for ships to use. People and goods travel on the rivers throughout Europe and to the open sea. Fast-flowing rivers also provide electricity.

Europe's Resources (pages 278–279)

Identifying

Identify 10 of Europe's natural resources.

1. _____
2. _____
3. _____
4. _____
5. _____
6. _____
7. _____
8. _____
9. _____
10. _____

Europe has many valuable natural resources. These resources have helped Europe become a leader in the world economy.

Energy Resources

Coal has been a key energy source for Europe. In the 1800s, coal fueled early factories. Today Europe supplies almost half of the world's coal. Many people in Europe work as coal miners.

Two other energy resources are natural gas and petroleum. Productive oil fields are found beneath the North Sea in areas controlled by Norway and the United Kingdom.

Europeans also use clean energy sources that cause less pollution. Swift-flowing rivers create hydroelectric power. Wind farms use turbines with fanlike blades to make electricity.

Other Natural Resources

Other resources include iron ore and manganese used to make steel. Marble and granite provide building materials. Many of Europe's once-vast forests have been cut down, however.

Fertile soil allows farmers to grow large amounts of crops, including rye, oats, wheat, and potatoes. Europe's waterways provide another valuable resource—fish.

Notes | Read to Learn

Environmental Issues (pages 279–280)

Problem-Solving

Describe how Europeans are attempting to solve each problem below.

1. Air pollution, acid rain

   _____
   _____
   _____

2. Water pollution

   _____
   _____
   _____

Air Pollution and Acid Rain

Smoke from burning oil and coal creates air pollution, which causes breathing problems and other health risks. When chemicals in air pollution mix with precipitation, acid rain results. Acid rain harms trees and damages the surfaces of buildings. Acid also builds up in lakes and rivers, poisoning fish and other wildlife.

Water Pollution

The waterways in and around Europe are polluted. Sewage, garbage, and industrial waste are dumped into the region's seas, rivers, and lakes. Chemicals in pesticides and fertilizers run off from farmland into rivers, harming fish and other marine life.

Finding Solutions

Europeans are working to fix environmental problems. Factories are trying to release fewer chemicals into the air. Lakes are being treated with lime to reduce acid rain damage. Waste and sewage are being treated to provide cleaner water. Farmers are using less fertilizer to reduce the amount of chemical runoff. More Europeans are recycling and reducing the amount of garbage they produce.

Section Wrap-Up *Answer these questions to check your understanding of the entire section.*

1. **Evaluating** How are waterways important to Europe?

   _____
   _____

2. **Determining Cause and Effect** How have Europe's resources contributed to environmental problems?

   _____
   _____

On a separate sheet of paper, write a paragraph comparing and contrasting the landforms in your area to the landforms found in Europe. Which are similar? Which are different?

Chapter 10, Section 1

Chapter 10, Section 2 (Pages 282–288)
Climate Regions

Big Idea

The physical environment affects how people live. As you read, complete the chart below by listing and describing Europe's main climate zones.

Climate Zone	Characteristics

Read to Learn

Wind and Water (page 283)

Finding the Main Idea

What is the main idea of this subsection?

_____
_____
_____
_____

Europe is farther north than much of the United States. Because of this, you might expect its climate to be cold. However, the North Atlantic Current carries warm water from the Gulf of Mexico to Europe. Winds from the west, called westerlies, pass over this current and also carry warmth to Europe.

Other wind patterns affect regions in Europe. Winds blowing north from Africa warm southern Europe. In contrast, winter winds from Asia can lower temperatures in eastern Europe.

The waters around Europe also influence its climate. Winds blowing from the seas help cool the land in summer. In winter, the same winds warm the cold land. Thus, coastal Europe has a more moderate climate than inland areas.

Climate Zones (pages 284–288)

Europe has three main climate zones—marine west coast, humid continental, and Mediterranean. Five additional climate zones appear in small areas of Europe—subarctic, tundra, highland, steppe, and humid subtropical.

Climate Zones (continued)

Identifying
Underline the types of vegetation for each climate zone.

Explaining
How do mountains affect southern Europe's climate?

_____
_____
_____
_____
_____
_____

Differentiating
Complete this sentence.

_____ winds are cold and dry, and _____ winds are hot and dry.

Marine West Coast
Much of northwestern and central Europe has a marine west coast climate. This climate features mild temperatures and much rain. The climate supports a long growing season. Rainfall occurs mostly in the autumn and early winter. Forests flourish in the marine west coast climate. Some forests have **deciduous** trees, which lose their leaves in the fall. In cooler areas of the zone, **coniferous** trees, also called evergreens, grow.

Humid Continental
The second main zone is the humid continental climate. It is found in eastern Europe and parts of northern Europe. This zone has cooler summers and colder winters than the marine west coast zone. It also gets less rain and snow. The humid continental zone has mixed forests of deciduous and coniferous trees.

Mediterranean
The Mediterranean zone is Europe's third major climate zone, and it includes much of southern Europe. Mediterranean summers are hot and dry, and winters are mild and wet. The Pyrenees and Alps stop cold winds from blowing into southern Europe. Southern France, however, experiences cold, dry, **mistral** winds from the north.

Hot, dry winds from Africa, known as **siroccos,** pick up moisture as they cross the Mediterranean Sea. They bring humid weather to much of southern Europe. Vegetation in the Mediterranean zone includes olive trees, grapes, and low-lying shrubs—plants that do not need much water.

Subarctic and Tundra
The cold subarctic zone covers parts of Norway, Sweden, and Finland. Evergreen trees grow in this zone. Even farther north is the tundra zone, a frigid area of treeless plains. Only low shrubs and mosses can grow in the tundra. During winter, the sun's rays reach this region for only four hours per day.

Highland
The highland climate zone is found in the high altitudes of the Alps and Carpathian Mountains. This climate is generally cool to cold. Temperature and rainfall vary in this zone, depending mainly on a place's altitude, or elevation above sea level. Evergreens grow part of the way up the mountains, stopping at a point called the timberline. Only scrubby bushes and low-lying plants grow above the timberline.

Chapter 10, Section 2

Notes | Read to Learn

Climate Zones (continued)

Listing

What are the five smaller climate zones in Europe?

1. _____
2. _____
3. _____
4. _____
5. _____

Other Climate Zones

The southern part of Ukraine is a dry, treeless grassland called a steppe. The climate—also called steppe—is dry too, but not as dry as a desert. An area of land north of the Adriatic Sea has a humid subtropical climate. Summers are hot and wet here, and winters are mild and wet.

Climate Change

Many European leaders are worried about global warming. They fear that melting glaciers will cause ocean levels to rise, flooding coastal areas of Europe. Millions of Europeans live on or near the coasts. Officials are encouraging people to change their patterns of energy use. Most European governments have also signed the Kyoto Treaty, which limits the output of greenhouse gases that cause global warming.

Section Wrap-Up

Answer these questions to check your understanding of the entire section.

1. **Defining** What are westerlies? How do they affect the climate of Europe?

   _____
   _____

2. **Specifying** In which of Europe's climate zones do coniferous, or evergreen, trees grow?

   _____
   _____

Imagine that you live in the tundra climate zone. Write a paragraph describing what a typical winter day might be like.

_____
_____
_____
_____
_____

Chapter 11, Section 1 (Pages 294–303)
History and Government

Big Idea

The characteristics and movement of people impact physical and human systems. As you read, complete the time line below with at least five key events and dates in Europe's history.

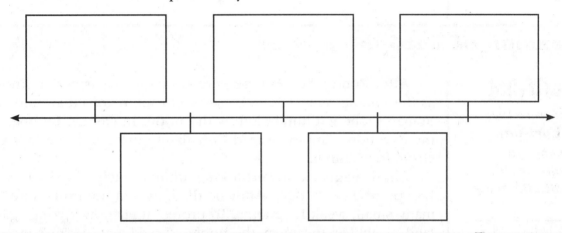

Notes | Read to Learn

Ancient Europe (pages 295–297)

Evaluating

How did ancient Greece and Rome influence later civilizations?

Ancient Greece:

_____

_____

_____

Ancient Rome:

_____

_____

_____

European civilizations began about 2,500 years ago in ancient Greece and Rome—known as the **classical** world. Greece's many mountains, islands, and seas separated early communities. Groups formed independent **city-states,** made up of the city and the area surrounding it. The city-state of Athens became the world's first **democracy,** in which citizens shared in running the government. Athens's democracy set an example for later civilizations. Learning and the arts also thrived.

Philip II of Macedonia conquered Greece in the mid-300s B.C. His son, Alexander the Great, made more conquests, spreading Greek culture to Egypt, Persia, and India. Rome conquered Greece in about 130 B.C.

Rome, located on the Italian Peninsula, became a **republic** in 509 B.C. In a republic, citizens choose their leaders. Rome also developed a code of laws known as the Twelve Tables. About 200 B.C., Roman armies began to conquer lands in the Mediterranean region. The republic grew into the massive Roman Empire. In A.D. 27, Augustus became Rome's first **emperor,** or all-powerful ruler. Christianity arose during his rule, and in A.D. 392, it became Rome's official religion.

 **Read to Learn**

Ancient Europe (continued)

Attacks from Germanic groups finally caused the Roman Empire to fall in A.D. 476. But Rome's influence was long lasting. Roman law shaped later legal systems. The language of Rome was Latin, the basis of many modern languages. Rome's architectural style became a model for Europe and the West.

Expansion of Europe (pages 298–300)

Identifying

1. What were the two Christian branches in Europe during the Middle Ages?

   _____
   _____

2. Which two religious groups fought in the Crusades?

   _____
   _____

Making Connections

Underline the phrase that explains why feudalism began. Then circle the phrase that explains why feudalism declined.

After Rome's fall, Europe entered a 1,000-year period known as the Middle Ages. Life in western Europe revolved around the Roman Catholic Church, led by the **pope.** In eastern Europe, the Byzantine Empire spread Christianity through the Eastern Orthodox Church.

Charlemagne, a powerful king, united much of western Europe in the A.D. 800s. After he died, his empire broke into many small, weak kingdoms. To protect their power, kings gave land to nobles. In return, the nobles served as the king's army in a system known as **feudalism.**

Christianity united much of Europe. But Muslims, or followers of Islam, spread through Southwest Asia and North Africa after the A.D. 600s. They also seized Palestine, which Christians viewed as their Holy Land. This led to religious wars called the Crusades. The Crusades opened trade with Muslim lands, which provided tax money for kings. Feudalism declined. Europe's kingdoms began to develop into **nation-states**—countries whose people shared a common culture or history.

The Renaissance, or "rebirth," was a time of renewed interest in art and learning from 1350 to 1550. The Renaissance thrived in Italian city-states like Florence, Rome, and Venice. Artists such as Michelangelo and Leonardo da Vinci created masterpieces. Humanism, or the belief in the importance of the individual rather than the Church, began to emerge.

In 1517 a religious leader named Martin Luther tried to reform, or change, some church practices he thought were wrong. The Reformation led to Protestantism, a new form of Christianity. Religious wars erupted throughout Europe. Monarchs gained power as church leaders lost theirs.

With new power and new technology, monarchs sent explorers overseas to look for spices and gold. Spain grew wealthy after Christopher Columbus landed in the Americas in 1492. Soon European monarchs began setting up colonies in the Americas, Asia, and Africa.

Modern Europe (pages 301–303)

Specifying

Underline and number three "revolutions" that swept Europe beginning in the 1600s.

Summarizing

What is the goal of the European Union?

_____

_____

Europe experienced several **revolutions,** or sweeping changes, beginning in the 1600s. During the Scientific Revolution, people used science and reason as a guide, rather than faith or tradition. The 1700s became known as the Age of Enlightenment. Thinkers such as John Locke said that all people have natural rights, including the rights to life, liberty, and property. Political revolutions in several countries limited the power of government.

During the Industrial Revolution, which began in Britain, machines and factories made goods faster and cheaper. Many Europeans left their farms to find work in cities.

Industry helped countries grow more powerful. They developed new weapons and competed for colonies. Tensions soon led to World War I (1914–1918) and World War II (1939–1945). Europe was left in ruins, and millions of people died. Six million Jews were killed in the **Holocaust.**

After World War II, the United States and Soviet Union fought for world power during the Cold War era. Much of Western Europe allied with the United States. Eastern European countries allied with the Communist Soviet Union. **Communism** is a system in which government controls the ways of producing goods. After the Soviet Union broke apart in 1991, many former Communist countries became democratic.

In 1993 several countries formed the European Union. Other countries have since joined. Their goal is to unify Europe. Workers and products move freely among member countries.

Section Wrap-Up

Answer these questions to check your understanding of the entire section.

1. **Explaining** What role did religion play in the history of Europe?

_____

_____

2. **Listing** What important changes occurred during the Middle Ages?

_____

_____

On a separate sheet of paper, write a paragraph describing one or more of the basic rights you have as an American citizen. Include ideas from the Enlightenment in your paragraph.

Chapter 11, Section 1

Chapter 11, Section 2 (Pages 306–312)
Cultures and Lifestyles

Big Idea

Europe is home to many different cultural groups. As you read, complete the diagram below by listing five key facts about Europe's population patterns.

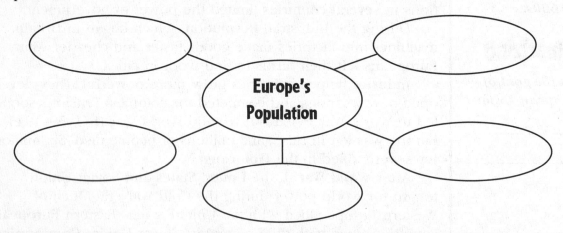

Notes	Read to Learn

Population Patterns (pages 307–308)

Determining Cause and Effect

What has caused Europe's ethnic mix?

_____
_____
_____

A Rich Ethnic Mix

Many different ethnic groups live in Europe. An **ethnic group** is a collection of people who share the same ancestry, language, and customs. Migrations, wars, and changing boundaries have led to Europe's ethnic mix. Ethnic loyalties create bonds. They also create conflict. Fighting among ethnic groups resulted in Yugoslavia splitting into five separate countries in 1990. Europeans are working toward unity, however. They value democracy and human rights. They expect their governments to care for citizens. Many European countries are **welfare states,** in which the government provides care for the sick, needy, and retired.

Population Changes

Europe's population continues to change. Since World War II, people have immigrated from Asia, Africa, and Latin America. Immigrants are not always welcomed. They compete with residents for jobs, housing, and other services. Some countries help immigrants adapt quickly to their new nation. Others pass laws to prevent immigrants from coming in at all.

Population Patterns (continued)

The continent's total population is declining. The reason is that Europe's **fertility rate**—the average number of children born to each woman—is low. Experts predict that by the year 2050, Europe will have 10 percent fewer people. Fewer people will mean fewer workers to keep the economy growing. In addition, young workers will have to pay higher taxes to support an increasingly older population.

Life in Europe (pages 308–310)

Listing

What are four ways to travel in Europe?

1. _____
2. _____
3. _____
4. _____

Explaining

Underline the sentence that explains why people in eastern European countries have lower incomes than those in western European countries.

Cities and Transportation

The Industrial Revolution in the late 1700s changed Europe from a rural, farming society to an urban, industrial one. **Urbanization** resulted, with most people living in cities and towns. Today Paris and London are two of the largest cities in the world.

Most European cities have government-owned public transportation systems. A vast rail system links cities and towns throughout Europe. France developed high-speed trains, which help protect the environment. Subways are common. Trains even speed through an underwater tunnel—called the Chunnel—between England and France. Highways also allow high-speed travel, particularly Germany's autobahn. Canals and rivers are widely used to ship goods, and ports dot Europe's long coastline. Airports also connect European cities.

Education, Income, and Leisure

Europeans are well educated and have a high literacy rate. As a result, many Europeans earn more money than people in other parts of the world. Service industries provide many jobs. However, incomes in northern and western Europe are higher than those in southern and eastern Europe. Many eastern European countries are still rebuilding from ethnic conflicts or from years under Communist rule.

The overall higher incomes mean that people have more money to spend on leisure activities. Europeans like to travel. France and Italy are popular vacation spots. Sports are popular, too. Ice hockey, skiing, rugby, and soccer are favorite pastimes.

Chapter 11, Section 2

Religion and the Arts (pages 310–312)

Sequencing

How did European art change from ancient times to the 1900s?

_____
_____
_____
_____
_____
_____
_____

Many Europeans are **secular,** or nonreligious. Yet religion has had a major impact on the life and art of Europe. Roman Catholicism is practiced in western Europe and some eastern European countries. Northern Europeans are mainly Protestant. Eastern Orthodoxy is practiced in the southern part of eastern Europe. Judaism and Islam have also influenced Europe's culture. For the most part, Europeans of different religions live together peacefully. But religious differences in Europe have sometimes led to violence.

Arts

The ancient Greeks and Romans built temples with huge, graceful columns. Gothic cathedrals with pointed arches and stained-glass windows arose during the Middle Ages. Art often focused on holy subjects or religious symbols. Renaissance artists and writers focused on religion too, but they also portrayed lifelike figures and believable characters in their works. In the 1800s, artists, writers, and musicians tried to stir emotions in a style known as Romanticism. Impressionists used bold colors to create "impressions" of the natural world. In the 1900s, painters expressed feelings and ideas in abstract paintings.

Section Wrap-Up *Answer these questions to check your understanding of the entire section.*

1. **Explaining** How do individual European countries deal with immigration?

   _____
   _____

2. **Drawing Conclusions** Why do most Europeans have higher incomes than people in other parts of the world?

   _____
   _____

On a separate sheet of paper, write a paragraph explaining some of the problems that Europeans will face as their population declines.

Chapter 12, Section 1 (Pages 320–328)
Northern Europe

Big Idea

Geographers organize the Earth into regions that share common characteristics. As you read, complete the chart below. List key facts about the people and cultures of northern Europe.

People and Cultures		
United Kingdom	**Ireland**	**Scandinavia**

Notes Read to Learn

The United Kingdom (pages 321–324)

Naming

What four regions make up the United Kingdom?

1. _____
2. _____
3. _____
4. _____

Identifying

What waterway separates Great Britain from mainland Europe?

_____

The United Kingdom is an island nation northwest of mainland Europe. England, Scotland, and Wales make up the island of Great Britain. Northern Ireland, also part of the United Kingdom, is located in one corner of the nearby island of Ireland. The English Channel separates Great Britain from the rest of Europe. Britain's southern and eastern plains have fertile farmland. The highlands and mountains of Scotland and Wales are best suited to sheep herding. London, the capital, is a world center of finance and business.

Great Britain led the Industrial Revolution, and it still exports manufactured goods. Electronics industries, banking, and health care also are vital to the economy. Oil and natural gas beneath the North Sea supply most of Britain's energy.

The United Kingdom is both a **constitutional monarchy** and a **parliamentary democracy.** A king or queen is the ceremonial head of state, but voters elect members of Parliament, the lawmaking body. The leader of the political party with the most members in Parliament becomes prime minister. Scotland, Wales, and Northern Ireland have regional legislatures with control over education and health care.

Chapter 12, Section 1   79

Notes | Read to Learn

The United Kingdom (continued)

Specifying

Circle the three languages spoken in Great Britain.

Great Britain has the third-largest population in Europe. Almost 90 percent of the people live in cities.

British people speak English. Welsh and Scottish Gaelic also are spoken in some areas. Most of the people are Protestant Christians, but a growing number of immigrants practice Islam, Hinduism, and Sikhism.

The United Kingdom was a powerful empire in the 1700s and 1800s. British culture—including the English language, the sport of cricket, and British literature—spread to many lands.

The Republic of Ireland (pages 325–326)

Defining

Define the word **productivity** *by using it in a sentence.*

_____
_____
_____
_____

The Republic of Ireland is an independent Catholic country. Nicknamed the Emerald Isle, Ireland is lush and green because of its regular rainfall. **Peat,** or plants partly decayed in water, is found in low-lying areas. The peat is dug from **bogs**—swampy lands. The peat is dried and can be burned for fuel.

Farmers raise sheep and cattle and grow sugar beets and potatoes. More people work in manufacturing than farming. Ireland's industries produce clothing, pharmaceuticals, and computer equipment. Increased **productivity**—how much work a person does in a set amount of time—has boosted the economy.

Celts settled the island hundreds of years ago. Ireland's two languages are Irish Gaelic (a Celtic language) and English. Dublin is the capital. Irish music and folk dancing are performed all over the world. Irish playwrights, novelists, and poets had a great influence on world literature.

Catholics in Northern Ireland would like to unite with their southern neighbors. Protestants there want to remain part of the United Kingdom, however. This dispute led to violence from the 1960s to the 1990s. A 1998 agreement has not halted the dispute.

Scandinavia (pages 326–328)

Naming

Underline the five countries that make up Scandinavia.

Scandinavia is the northernmost region of Europe. It includes five countries—Norway, Sweden, Finland, Denmark, and Iceland. Their standards of living are among the highest in the world.

The northernmost part of Scandinavia is always cold. The southern and western areas are mild due to the warm North Atlantic Current. Islands dot the jagged coastline. Lowland plains cover Denmark and southern Sweden and Finland. Mountains rise

Scandinavia (continued)

Listing

List seven features of Scandinavia's landscape.

1. _____
2. _____
3. _____
4. _____
5. _____
6. _____
7. _____

along Norway and Sweden's shared border. Forests and lakes cover Sweden and Finland. Barren, frozen tundra is found above the Arctic Circle. In Iceland, tectonic activity creates springs called **geysers** that shoot hot water and steam into the air. Norway has many narrow sea inlets called **fjords,** which are surrounded by steep cliffs carved by glaciers.

The strong economies of Scandinavia consist of agriculture, manufacturing, and service industries. Fishing also is important. Each country uses different sources of energy. Norway relies on oil and natural gas pumped from its fields under the North Sea. Iceland uses **geothermal energy,** or electricity produced by underground steam. Finland generates hydroelectric power from its fast-running rivers. Sweden uses nuclear power and oil.

Norway, Sweden, Denmark, and Iceland share ethnic ties and speak similar languages. Finland's language and culture are different, yet it shares close historic and religious links to the other countries. Most people are Protestant Lutherans. Denmark, Norway, and Sweden are constitutional monarchies like the United Kingdom. Finland and Iceland are republics with elected presidents. All five countries are welfare states. They provide health care, child care, elder care, and retirement benefits to all. In return for these services, citizens pay some of the highest taxes in the world.

Section Wrap-Up *Answer these questions to check your understanding of the entire section.*

1. **Differentiating** What are some of the differences between Northern Ireland and the Republic of Ireland?

   _____
   _____

2. **Explaining** What is a welfare state?

   _____
   _____

On a separate sheet of paper, write a paragraph explaining how the government of the United Kingdom is different from the government of the United States.

Chapter 12, Section 1

Chapter 12, Section 2 (Pages 329–337)
Europe's Heartland

Big Idea

People's actions can change the physical environment. As you read, complete the Venn diagram below. Compare and contrast France and Germany.

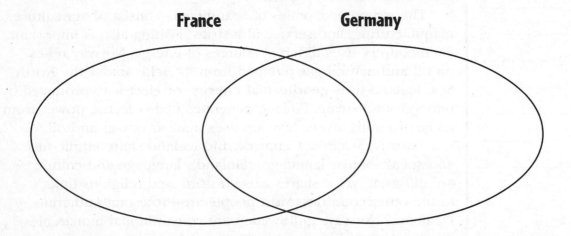

Read to Learn

France and the Benelux Countries (pages 330–333)

Specifying

What products do French farmers specialize in?

_____

_____

Stating

What are two landmarks in Paris?

_____

_____

France lies in the western part of Europe. Its mild climate and rich soil are good for farming. French farms are known for their **specialization,** or focusing on using their best resources. Some farms grow grapes for wine, while other farms produce cheese. Traditional industries make cars and trucks, chemicals, and textiles. **High-technology industries** make computers and other products that require specialized engineering. Tourism is another major part of the French economy. Millions of tourists visit Paris, the capital, and its museums and cathedrals, such as the Louvre and Notre Dame. They also tour the country's historic castles.

Most people speak French and are Roman Catholic. Immigration from Muslim countries has made Islam France's second-largest religion. The majority of people live in urban areas.

The French are proud of their culture. French cooking and fashion are admired worldwide. France also boasts famous writers, philosophers, artists, and composers. France is a democratic republic with an elected president and a prime minister appointed by the president.

France and the Benelux Countries (continued)

Naming

What are the three Benelux countries?

_____

_____

_____

Explaining

How have the Dutch increased their farmland?

_____

_____

The Benelux Countries

Belgium, Netherlands, and Luxembourg—also known as the Benelux countries—have a low, flat landscape. Most people live in crowded cities, work in businesses or factories, and have a high standard of living. All three countries are parliamentary democracies with monarchs.

Belgium has three distinct cultural regions: Flanders has Dutch-speaking Belgians, Wallonia has French-speaking Belgians, and the Brussels region is **bilingual,** or has two official languages. Brussels is the capital and headquarters of the European Union (EU).

The people of the Netherlands are known as the Dutch. They have built dikes and drained land that once was covered by the sea, turning it into rich farmland called **polders.** The Dutch make good use of their limited space, building narrow but tall homes. Amsterdam is the capital and largest city.

Luxembourg, a center of trade and finance, is the headquarters for many **multinational companies,** or firms that do business in several countries. The people have a mixed French and German background.

Germany and the Alpine Countries (pages 334–337)

Listing

List four things that Germany produces.

1. _____
2. _____
3. _____
4. _____

Germany has flat plains in the north, rocky highlands in the center, and the Alps in the far south. The Danube, Elbe, and Rhine Rivers are used to transport raw materials to factories and finished goods to markets.

Separate states made up the region for centuries. In 1871 these states joined to create the country of Germany. In the early 1900s, Germany's attempts to control Europe led to two world wars. In 1945 the Soviet Union controlled Communist East Germany. West Germany became democratic. **Reunification** came about in 1990, when the two parts of Germany were united into one country. Today Germany is a federal republic. A chancellor chosen by parliament is head of the government.

Germany has the largest population in Europe. Berlin is Germany's capital and largest city. Nearly everyone speaks German and is Protestant or Catholic. Germany is a global economic power and a leader in the European Union. The country grows enough food to feed its people and to export to other countries. Industry drives the strong economy, however. Factories produce steel, chemicals, cars, and electrical equipment.

Chapter 12, Section 2

Notes | Read to Learn

Germany and the Alpine Countries (continued)

Identifying
What are Switzerland's four official languages?

1. _____
2. _____
3. _____
4. _____

Speculating
Why do you think people consider Swiss banks secure?

_____
_____
_____
_____

The Alpine Countries

The Alps are a mountain range in central Europe. The countries in this region—Switzerland, Austria, and Liechtenstein—are known as the Alpine countries. Liechtenstein is the smallest, with a population of about 40,000.

Landlocked Switzerland practices **neutrality,** meaning it does not take sides in wars. Its stable democracy has made Switzerland a home to many international organizations, such as the International Red Cross. The rugged Alps separated Switzerland's people, resulting in diverse traditions, ethnic groups, and religions. The four official languages of Switzerland are German, French, Italian, and Romansch. Many Swiss speak more than one language.

Switzerland is a thriving industrial nation. Hydroelectricity powers its industries and homes. Swiss workers make electrical equipment, chemicals, watches, chocolate, and cheeses. People around the world consider Swiss banks safe and secure, so financial services make up a large part of Switzerland's economy.

Austria is a landlocked country east of Switzerland. The Alps cover most of the country and provide timber, iron ore, and beautiful scenery for tourists. Hydroelectricity powers Austria's factories, which produce machinery, chemicals, metals, and vehicles.

Most Austrians live in cities and towns, speak German, and are Roman Catholic. The capital, Vienna, is a center of culture. The city is known for its concert halls, palaces, and churches.

Section Wrap-Up
Answer these questions to check your understanding of the entire section.

1. **Sequencing** List three key events and dates in Germany's recent history on the time line below.

   _____|_____|_____|_____

2. **Explaining** What are the Alpine countries, and why are they called that?

   _____
   _____

Do you think it is important for a person to know more than one language? On a separate sheet of paper, write a paragraph presenting your opinion and persuading your readers to agree.

84   Chapter 12, Section 2

Chapter 12, Section 3 (Pages 338–342)
Southern Europe

Big Idea

Places reflect the relationship between humans and the physical environment. As you read, complete the diagram below. Write three characteristics shared by Spain, Italy, and Greece

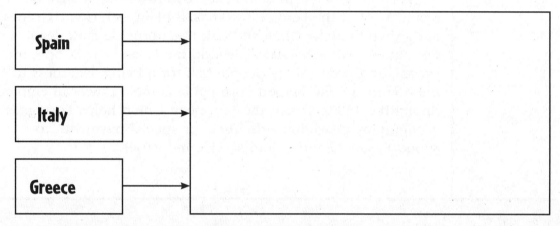

Notes | Read to Learn

Spain and Portugal (pages 339–340)

What are Spain and Portugal's chief agricultural products?

Spain:

_____

_____

_____

Portugal:

_____

_____

Spain and Portugal are located on the Iberian Peninsula. A third country, tiny Andorra, is nestled in the Pyrenees Mountains. The Meseta, a dry plateau surrounded by mountains, covers most of Spain. Because little rain falls on the Meseta, farmers use a technique called **dry farming** to grow wheat and vegetables. The land is left unplanted every few years so it can store moisture. The milder climate of southern Spain is good for growing citrus fruits, olives, and grapes.

Manufacturing and service industries dominate the economy. Spain produces foods, clothing, footwear, steel, and cars. Tourism also boosts the economy. Tourists visit Spain's castles, cathedrals, and Mediterranean beaches. They also enjoy Spanish bullfighting and flamenco dancing.

Castilian Spanish is Spain's official language, but people speak different languages in the country's various regions. The people of Catalonia speak Catalon. The Basques speak Euskera, a language unrelated to any other language in the world. Spain's democratic government has given the different regions a great deal of **autonomy,** or self-rule. In the Basque region, many people want to be independent.

Notes | Read to Learn

Spain and Portugal (continued)

Locating

Where do most Portuguese live?

_____
_____
_____
_____

Spain's main cities are Madrid, the capital, and Barcelona, the leading seaport and industrial center. Most people live in cities and are Roman Catholic, but a large number of Muslims have immigrated.

Portugal

Portugal is a small, democratic country located west of Spain. Most of the land, a low coastal plain, is farmed. Grapes are grown to make wine, and oak trees provide cork. Most Portuguese live in small villages on the coast near Lisbon, the capital, and Porto. Many people fish for a living. Nearness to the Atlantic Ocean helped Portugal become a powerful empire during the 1500s. Today, the European Union helps Portugal's economy by providing **subsidies,** or special payments, to support manufacturing and service industries.

Italy (page 341)

Explaining

Explain why northern Italy is richer than southern Italy.

_____
_____
_____
_____

Describing

What and where is Vatican City?

_____
_____
_____
_____

Italy is a boot-shaped peninsula that extends into the Mediterranean Sea. The Alps dominate northern Italy. The Apennine Mountains run through the central and southern areas, including the Italian island of Sicily. Volcanoes are found throughout the country.

Italy has an industrial economy. Most industry is based in northern Italy in the cities of Milan, Turin, and Genoa. Workers there produce cars, technical instruments, appliances, and clothing. The Po River valley in the north also is a rich farming region where farmers raise livestock and grow grapes, olives, and other crops.

The mountainous land of southern Italy is not good for farming and is poorer than northern Italy. Many people have left the region because it has less industry and higher unemployment.

Italy is a democratic republic. Its people speak Italian, and nearly all are Roman Catholic. About 90 percent live in urban areas. Rome, the capital and largest city, once was the center of the Roman Empire. The Roman Catholic Church is based in Rome. The Church rules Vatican City, which is actually an independent country within Rome's boundaries.

Greece (page 342)

Listing

List five components of Greece's economy.

1. _____
2. _____
3. _____
4. _____
5. _____

Identifying

What is the Parthenon?

_____

Greece lies east of Italy and extends from the Balkan Peninsula into the Mediterranean Sea. In addition to its mainland, Greece also has about 2,000 islands. Mountains and poor, stony soil cover Greece, so agriculture is not as important to the economy. People do raise sheep and goats in the highlands, however, and farmers grow wheat and olives in valleys and plains.

The textile, footwear, and chemical industries have become important to Greece's economy in the last few decades. Shipping and tourism also are vital to the country's economy. Greece boasts a large shipping fleet, including oil tankers, cargo ships, and passenger ships. Millions of visitors come each year to see historic sites like the Parthenon, an ancient temple in the city of Athens.

Almost 60 percent of the people of Greece live in urban areas, with one-third of them living in or near Athens, the capital. The people speak Greek, and most follow the Greek Orthodox Christian religion. Greece is a member of the European Union, and it has a democratic republic form of government.

Section Wrap-Up

Answer these questions to check your understanding of the entire section.

1. **Explaining** Why has membership in the European Union been good for Portugal?

   _____
   _____

2. **Sequencing** List southern Europe's peninsulas in order from west to east.

   _____
   _____

Imagine that you are a reporter assigned to interview a person who has just returned from a trip to southern Europe. Create an outline of important questions to ask in your interview.

_____
_____
_____
_____

Chapter 12, Section 4 (Pages 348–356)
Eastern Europe

Big Idea

Geography is used to interpret the past, understand the present, and plan for the future. As you read, complete the chart below. Summarize key facts about each group of countries.

Country Groups	Key Facts
Poland, Belarus, Baltic Republics	
Czech Republic, Slovakia, Hungary	
Countries of South-eastern Europe	

Read to Learn

Poland, Belarus, and the Baltic Republics (pages 349–350)

Identifying

Complete these sentences with the names of the countries they describe.

A flat landscape made

_____

easy to invade.

_____

has a command economy.

Poland borders the Baltic Sea. Fertile lowland plains cover most of the country. The Carpathian Mountains rise in the south and west. Throughout history, Poland's flat landscape made it easy for other countries to invade. In 1939 Germany attacked Poland, starting World War II. After the war, a Communist government ruled Poland and set up a **command economy**—the government decided what, how, and for whom goods would be made. Factories made military goods instead of food, which led to shortages. Polish workers and farmers formed Solidarity, a labor group that pushed for democracy and a better life. Communist leaders allowed free elections in 1989, and Poland became a democracy. This event led to the fall of other Communist governments in Eastern Europe. Today Poland has a **market economy**—people and businesses decide what to produce. Agriculture is still important, but industries are growing. Warsaw is the capital.

Belarus, east of Poland, also is covered by a lowland plain. A former Soviet republic, Belarus now has a rigid government and a command economy. Its main resource is **potash**, a mineral used to make fertilizer. Government-controlled farms grow

88   Chapter 12, Section 4

Poland, Belarus, and the Baltic Republics (continued)

Naming

The Baltic Republics are _____, _____, and _____.

grains, and factories make trucks, radios, TVs, and bicycles. Most people are Eastern Orthodox Slavs.

The Baltic Republics—Lithuania, Latvia, and Estonia—were under Russian control until the Soviet Union broke apart in 1991. Now they are democracies with strong economies based on dairy farming, beef production, fishing, and shipbuilding.

The Czech Republic, Slovakia, and Hungary (pages 352–353)

Listing

List six items produced by the Czechs.

_____
_____
_____
_____
_____
_____

Specifying

From whom did the Hungarians descend?

_____

The Czech Republic, Slovakia, and Hungary are landlocked countries. The Czech Republic and Slovakia were one country, Czechoslovakia, from 1918 to 1993. Once under Communist rule, all three countries now have democratic governments.

The Czech Republic features rolling hills, lowlands, and plains bordered by mountains. Prague, the capital, is known for its historic buildings. Czechs enjoy a relatively high standard of living. The country is a major agricultural producer, but manufacturing is the core of its economy. Goods include machinery, vehicles, metals, and textiles, as well as fine crystal and beer.

The Carpathian Mountains tower over northern Slovakia with rugged peaks, thick forests, and lakes. Vineyards and farms dot the south's fertile lowlands. Slovakia has been slow to move to a market economy. It has less industry than the Czech Republic. Most Slovaks are devout Catholics. Bratislava, the capital, is located on the Danube River.

Hungary's landscape is a lowland area with fertile farmland. Its capital, Budapest, straddles the Danube River. Hungary has few natural resources, but it imports raw materials for industry. The country exports chemicals, food, and other products. The Hungarians have a unique language and are descended from the Magyars of Central Asia.

Countries of Southeastern Europe (pages 353–356)

Stating

What physical feature divides Ukraine?

_____

Eleven countries in southeastern Europe are located along the Black Sea or Balkan Peninsula. Ukraine, the largest country in Europe, is located on a lowland plain with the Carpathian Mountains to its southwest. The Dnieper River divides the country into two sections. The lowland steppes to the west have rich soil good for farming and raising cattle and sheep. People of Ukrainian descent live in this "breadbasket of Europe," and they

Chapter 12, Section 4

89

Countries of Southeastern Europe (continued)

Summarizing

List two facts about western Ukraine and two facts about eastern Ukraine.

Western Ukraine:
_____
_____

Eastern Ukraine:
_____
_____

Identifying

What are Bulgaria's major economic activities?
_____
_____

want to join the European Union. Coal and iron ore are mined in the eastern plains, which are heavily industrialized. The people in this section are ethnic Russians who want closer ties to Russia. Ethnic divisions have grown sharp.

Romania has coal, petroleum, and natural gas, which helps its economy. Romans once ruled this region, so the Romanian language is based on the Latin spoken in ancient Rome. Bucharest, the capital, is the country's major commercial center.

Moldova is a landlocked country between Ukraine and Romania. It has fertile soil and productive farms but few mineral resources or industries. It is Europe's poorest country.

Bulgaria is a mountainous country with fertile river valleys that are good for farming. Many people work in factories located in Sofia, the capital, or in the tourism industry at Black Sea resorts.

Most countries on the Balkan Peninsula once were part of a Communist country called Yugoslavia. In the early 1990s, Slovenia, Croatia, Bosnia and Herzegovina, and Macedonia declared their independence. Serbia wanted to keep Yugoslavia together under Serbian rule, however. Serbs carried out **ethnic cleansing**, removing or killing entire ethnic groups. The conflicts left the Balkan countries scarred and with even poorer economies than they had under Communist rule. Their mountainous landscape is not good for farming, and they have few natural resources.

Albania is unique in that it is the only European country with a majority Muslim population. Albanian farmers outnumber factory workers, and this country too is poor.

Section Wrap-Up *Answer these questions to check your understanding of the entire section.*

1. **Explaining** What was Solidarity, and what was its goal?
_____
_____

2. **Sequencing** What caused Serbia to implement ethnic cleansing?
_____
_____

On a separate sheet of paper, discuss the differences between a command economy and a market economy. Explain why you think countries like Poland wanted to move to a market economy.

Chapter 13, Section 1 (Pages 372–375)
Physical Features

Big Idea

Changes occur in the use and importance of natural resources. As you read, complete the diagram below. List six of Russia's major landforms.

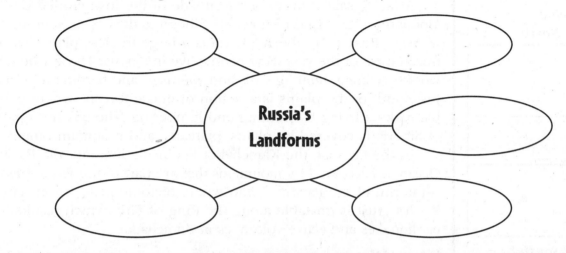

Notes | Read to Learn

Landforms of Russia (pages 373–374)

Naming

Name the two continents that Russia straddles and the mountain range that divides them.

Continents:

_____

_____

Mountain range:

_____

Russia is the largest country in the world, measuring about 6,200 miles from east to west. The country is located on two continents—Europe and Asia. The Ural Mountains divide the European and Asian parts of Russia.

Because the country is so large, Russia has a long coastline. Russia is located in the north, however, and the water along most of its coast is frozen for much of the year. Few ports in Russia are always free of ice. In the southwest, though, Russia does have port cities on the Black Sea. The Black Sea provides a warm-water route for ships traveling from inland Russia to the Mediterranean Sea.

European Russia

A variety of landforms cover Russia. Most of the European portion of Russia lies on the Northern European Plain. This fertile area has a mild climate, and about 75 percent of Russia's population lives there. Moscow, the capital, and St. Petersburg, a large port city near the Baltic Sea, are located on the Northern European Plain. Good farmland and grassy plains are found farther south along the Volga and other rivers.

Chapter 13, Section 1   91

Landforms of Russia (continued)

Describing

Describe the physical characteristics of the three sections of Siberia.

North

South

Specifying

Why is the Caspian Sea important to Russia?

_____
_____
_____

The rugged Caucasus Mountains rise to the far south in European Russia. Because the mountains are located along a fault line, destructive earthquakes strike the Caucasus area.

Asian Russia

Asian Russia is on the eastern side of the low, eroded Ural Mountains. The region known as Siberia makes up a huge part of Asian Russia. Northern Siberia is a large treeless plain that is frozen most of the year. Few people live in Siberia. Those who do work by fishing, hunting seals and walruses, and herding reindeer. Just south of the plains is a region of dense forests, where people make a living by hunting and lumbering. The southern part of Siberia is covered by plains, plateaus, and mountain ranges.

In the far east, the Kamchatka Peninsula juts into the Pacific Ocean. It is covered by mountains that are part of the Ring of Fire—the rim of the Pacific Ocean where tectonic plates meet. The Earth's crust is unstable along the Ring of Fire, which results in earthquakes and active volcanoes in Kamchatka.

Inland Waters

Russia has many inland waterways. The major river in European Russia is the Volga. In Siberia, many rivers begin in the southern mountains and flow north to the Arctic Ocean. The Caspian Sea, located in southwestern Russia, is the largest inland body of water in the world. A saltwater lake, it is important for its fish as well as its oil and natural gas deposits. Lake Baikal, the world's deepest freshwater lake, is located in southern Siberia. It holds one-fifth of the world's unfrozen freshwater. Baikal, or nerpa, seals and many other forms of aquatic life are found in the deep waters of Lake Baikal.

Natural Resources (page 375)

Identifying

Circle seven natural resources found in Russia.

Russia is filled with natural resources. The country has large reserves of **fossil fuels**—oil, natural gas, and coal. It also has major deposits of iron ore, copper, and gold. The iron ore has been used to build up Russia's steel industry.

Another valuable resource found in Russia is timber. Much of Siberia is covered by trees. Russia provides about 20 percent of the world's softwood. **Softwood,** the wood from evergreen trees, is used in buildings and for making furniture.

Many of Russia's resources are located in the frozen region of Siberia. The immense size and remote location of Siberia

 # Read to Learn

Natural Resources (continued)

Summarizing

What challenges does Russia face in accessing its resources?

_____
_____
_____

provide challenges to Russians who attempt to access the region's resources. Siberia lacks the **infrastructure,** or roads and railroads, needed to transport materials. The intense cold makes it difficult for workers to stay warm and to keep their equipment from freezing. Recently, a pipeline was built to carry natural gas from Siberia to Europe.

Section Wrap-Up

Answer these questions to check your understanding of the entire section.

1. **Explaining** Why does Russia not benefit from having a long coastline?

_____
_____

2. **Making Generalizations** Where do most Russians live? Why?

_____
_____

 *In the space provided, write a paragraph describing which region of Russia would be the most difficult to live in based upon physical characteristics.*

_____
_____
_____
_____
_____
_____
_____
_____

Chapter 13, Section 1

Chapter 13, Section 2 (Pages 378–382)
Climate and the Environment

Big Idea

People's actions change the physical environment. As you read, complete the diagram below. List four factors that lead to Russia's cold climate, especially those related to location and landforms.

```
[  ]        [  ]
   [  ]   [  ]
   Russia's Climate
```

 **Read to Learn**

A Cold Climate (pages 379–380)

Listing

List three climate zones that are found in Russia.

1. _____
2. _____
3. _____

Most of Russia is located in the high latitudes. Because it is so far north, Russia does not receive much of the sun's heat, even during the summer. Much of Russia also is inland, away from the oceans. In other parts of the world, the warm currents of the Atlantic Ocean and Pacific Ocean help moderate the temperature. Russia does not have this benefit.

The landforms in Russia also affect the country's climate. In the north, the elevation of the land is not high enough to block the cold, icy air that blows south from the Arctic region. In the south and east, tall mountains block the warm air that would otherwise come from the lower latitudes. As a result, most of Russia experiences only two seasons—long winters and short summers. Spring and autumn are short periods of changing weather.

Most of western Russia has a humid continental climate, with warm, rainy summers and cold, snowy winters. The average July temperature in Moscow is just 66°F. The average January temperature, in contrast, is 16°F. These cold winters have played an important role in Russia's history. Germany's advance into Russia during World War II was stopped by the bitter cold.

A Cold Climate (continued)

Speculating

What is a coniferous tree?

_____

In the northern and eastern parts of Russia, the summers are short and cool, and the winters are long and snowy. In the far north, the tundra climate zone has resulted in a permanently frozen layer of soil beneath the surface, called **permafrost.** Only mosses, lichens, and small shrubs survive in the tundra. The subarctic climate zone is located south of the tundra. It has slightly warmer temperatures than the tundra zone does. The **taiga**—the world's largest coniferous forest—stretches about 4,000 miles across Russia's subarctic region. The taiga is about the size of the United States.

Russia's Environment (pages 380–382)

Finding the Main Idea

What is the main idea of this subsection?

_____
_____
_____
_____
_____

Paraphrasing

Complete this sentence.

Other countries are providing Russia with _____ to improve _____ and clean up _____.

Throughout the 1900s, Russia's leaders focused on expanding the country's economy. In doing so, they paid no attention to how this growth damaged the environment.

Factories continue to release pollutants into the air today. **Pollutants** are chemicals and smoke particles that cause pollution. A thick haze of fog and chemicals, called **smog,** hovers over many of Russia's cities. As a result, a large number of Russians have lung diseases and cancer.

Water Pollution

Russia also has water pollution. Agriculture and industry use many chemicals. These chemicals often drain into waterways. Poor sewer systems are another source of water pollution.

One effect of water pollution is that some of Russia's animal species are threatened. For example, animal populations around Lake Baikal may be getting smaller because pollution has damaged the water. Water pollution also affects people. More than half of the people in Russia do not have safe drinking water.

Cleaning Up

Russia is making efforts to clean up its environment. Other countries are providing aid to Russia. Russia is using this assistance to improve the country's sewage systems. International aid also is helping Russia clean up heavily polluted sites.

Cities are building power plants that are more efficient. These new plants will use less energy than the existing ones. They also will burn fuel more cleanly. As a result, fewer pollutants will be released into the air. Some of these efforts are cleaning up the existing pollution, and others are intended to reduce future pollution. Even so, it will take a long time for Russia to have a healthy environment.

Chapter 13, Section 2

Section Wrap-Up

Answer these questions to check your understanding of the entire section.

1. **Determining Cause and Effect** What effects do air and water pollution have on people and animal life in Russia?

   _____

   _____

2. **Specifying** What are three steps Russia is taking to improve its environment? Which steps address existing pollution and which reduce future pollution?

   _____

   _____

   _____

   _____

Expository Writing

In the space provided, write a paragraph providing several suggestions on how a country can expand its economy without harming the environment.

_____

_____

_____

_____

_____

_____

_____

_____

_____

_____

_____

Chapter 14, Section 1 (Pages 388–394)
History and Governments

Big Idea

The characteristics and movement of people impact physical and human systems. As you read, make an outline of the section using the model below. Use Roman numerals to number the main headings. Use capital letters to list two key facts below each main heading.

I. First Main Heading
   A. Key Fact 1
   B. Key Fact 2
II. Second Main Heading
   A. Key Fact 1
   B. Key Fact 2

Notes | Read to Learn

The Russian Empire (pages 389–391)

Sequencing

Write down important dates and events in Russia's development.

_____
_____
_____
_____
_____
_____

Russia began as a small trade center. Slavic people lived along the rivers in Ukraine and Russia. In the A.D. 800s, Slavs settled the town of Kiev, which became the civilization known as Kievan Rus. **Missionaries,** or people who move to another area to spread their religion, brought Eastern Orthodox Christianity and a written language to Kievan Rus in A.D. 988.

In the 1200s, Mongol warriors from Central Asia conquered Kievan Rus. Many of the Slavic people moved north. They built a small trade center called Moscow. It became the center of a new Slavic territory called Muscovy. Ivan III, a prince of Muscovy, declared independence from Mongol rule. A strong ruler, he was known as "Ivan the Great."

In 1547 Ivan IV declared himself **czar,** or emperor, of Muscovy, which became Russia. Known as "Ivan the Terrible," he expanded his empire by conquering neighboring lands. Later czars, such as Peter the Great and Catherine the Great, also expanded the empire. They wanted a warm-water port for trade. They also wanted to become more European. Peter the Great built a new capital—St. Petersburg—close to Europe. The Russian Empire extended to the Pacific Ocean and Central Asia.

Chapter 14, Section 1   97

The Russian Empire (continued)

Explaining

How did Russia's physical geography act as a weapon in 1812?

_____

_____

_____

_____

The czars, large landowners, and wealthy merchants lived comfortably. The majority of Russians were serfs, however. **Serfs,** or peasant farm laborers, could be bought and sold with the land.

Russia's cold climate and huge size helped defeat invaders. The French emperor Napoleon invaded in 1812, but he lost most of his army retreating during the brutal Russian winter.

Russia changed greatly in the late 1800s. Czar Alexander II freed the serfs in 1861. He built industries and railroads to modernize Russia's economy. Yet most Russians remained poor, and unrest spread. In early 1917, the people revolted and overthrew Czar Nicholas II. Vladimir Lenin established a **Communist state** in which the government controlled the economy and society.

The Rise and Fall of Communism (pages 391–394)

Specifying

Underline the reason Lenin ended private ownership.

Circle the phrase that describes why factory goods were of poor quality.

Defining

What was the Cold War?

_____

_____

_____

_____

Lenin created a new country called the Union of Soviet Socialist Republics (U.S.S.R.), or the Soviet Union. It included 15 republics and many different ethnic groups. Lenin followed the ideas of Karl Marx, a German political thinker. Marx believed that industrialization was unfair because factory owners had much power while the workers had little. Lenin wanted to make everyone equal. He ended private ownership. The government took control of all factories and farms.

Later Soviet leaders, such as dictator Joseph Stalin, set up a command economy in which the government made all economic decisions. The government introduced **collectivization,** a system in which small farms were combined into larger ones. The collective farms were not efficient, however, and did not produce enough food for all the people. Soviet factories produced steel, machines, and military equipment. Without competition, however, many goods were of poor quality.

During World War II, Germany invaded the Soviet Union. About 20 to 30 million Russian soldiers and civilians died. After the war, Stalin kept troops in neighboring countries to make sure the U.S.S.R. would not be invaded again. He set up Communist governments in Eastern Europe.

Although the Soviet Union and the United States were allies during World War II, they became enemies after the war. Because no physical combat occurred, their conflict became known as the **Cold War.** The United States led democracies in Western Europe, and the Soviet Union led Communist Eastern Europe. Many Soviet resources were used to produce weapons.

Chapter 14, Section 1

 | **Read to Learn**

The Rise and Fall of Communism (continued)

Identifying

Who was the last president of the Soviet Union?

_____

Who were the first two presidents of an independent Russia?

_____

Mikhail Gorbachev became the Soviet leader in 1985. The people were tired of enduring shortages of food and other goods. Gorbachev established the policy of **glasnost,** or "openness." People could say or write their opinions without being punished. He also introduced **perestroika,** or "rebuilding," to boost the economy. Perestroika allowed for small, privately owned businesses. Gorbachev thought these new policies would strengthen the people's support of the government. Instead, Eastern Europeans began to doubt communism, and protests arose. By 1991, all of Eastern Europe's Communist governments had changed to democracies.

Hard-liners in the Soviet Union wanted to stop the changes and return to communism. Boris Yeltsin became president of Russia, the largest of the Soviet republics. In 1991 hard-liners attempted a **coup** to overthrow him by military force. The coup failed. Russia and all the other Soviet republics declared independence. By the end of 1991, the Soviet Union no longer existed as a nation.

Yeltsin worked to build a democracy and to create a market economy. But Vladimir Putin, who became president of Russia after Yeltsin, increased government controls to deal with rising crime and violence. Ethnic minorities, such as those in the Chechnya region, have tried to separate from Russia.

Section Wrap-Up *Answer these questions to check your understanding of the entire section.*

1. **Determining Cause and Effect** What effect did the command economy have on the Soviet Union, particularly during the Cold War?

_____

_____

2. **Explaining** What impact did Gorbachev's policies have on Eastern Europe?

_____

_____

 *On a separate sheet of paper, write a summary of how Russia started as a small trade center and grew to become the Russian Empire.*

Chapter 14, Section 1

Chapter 14, Section 2 (Pages 396–400)
Cultures and Lifestyles

Big Idea
Culture groups shape human systems. As you read, complete the diagram below by listing six details about the arts in Russia.

 **Read to Learn**

Russia's Cultures (pages 397–398)

Identifying

What are the three largest ethnic groups in Russia?

_____

_____

_____

Russia is home to dozens of different ethnic groups. The largest ethnic group includes Russians, or Slavs who descended from the people of Muscovy. The next-largest groups are Tatars, or Muslim descendants of Mongols, and Ukrainians, whose ancestors were the Slavs who settled around Kiev. Many of these groups speak their own language. Most people also speak Russian, the country's official language.

The Russian people were not allowed to practice religion during Communist rule, but now the people have religious freedom. The country's major religion is Eastern Orthodox Christianity. Many Muslims live in the Caucasus region.

The Arts
The arts have always been a central part of Russia's culture. The people had a strong **oral tradition,** meaning they passed on stories by word of mouth. Writers and musicians used these stories or folk music to create new works. One theme in Russian artistic works is **nationalism,** or feelings of loyalty toward the country.

100   Chapter 14, Section 2

Russia's Cultures (continued)

Naming
Who are two famous Russian writers?

_____

_____

Russia is a center of music and dance. Peter Ilich Tchaikovsky wrote the famous ballets *Swan Lake* and *The Nutcracker*. Two world-famous ballet companies are the Bolshoi in Moscow and the Kirov in St. Petersburg. Russia is also known for its literature, featuring such writers as Leo Tolstoy during the 1800s and Alexander Solzhenitsyn in the Communist era. The Hermitage Museum in St. Petersburg holds many masterpieces. There you can see the czars' jewel-encrusted Easter eggs made by Peter Carl Fabergé.

Specifying
What has been the major area of focus for Russian scientists?

_____

Scientific Advances
Because of its emphasis on science education, Russia has many scientists, mathematicians, and doctors. Some of the scientists' most significant work has been in the area of space exploration. In 1961 Russian Yuri Gagarin became the first person to fly in space.

Life in Russia (pages 399–400)

Listing
What are three popular sports and three national holidays in Russia?

Sports

_____

_____

_____

Holidays

_____

_____

_____

Russia is modernizing after decades of Soviet control. However, it also faces several challenges. Most of Russia's cities are west of the Ural Mountains. The majority of people in these cities live in large apartment buildings. Housing is scarce and expensive, so grandparents, parents, and children often live together. Wealthier people have country homes, called dachas. It is common for people to have vegetable gardens at their dachas. They either eat the vegetables or sell them in the cities.

In many areas of Russia, homes are designed to protect against the extreme cold. In Siberia, for example, some houses have three doors at the entrance. This keeps cold air from coming in when the outside door is opened.

Sports and Holidays
The most popular sports in Russia are winter sports or sports that are played indoors. Russian athletes are among the world's best in hockey, figure skating, and gymnastics.

Russians celebrate several national holidays. The newest holiday is Independence Day. It occurs on June 12, and it celebrates Russia's declaration of **autonomy,** or independence, from the Soviet Union. Another festive holiday is New Year's Eve. In the spring, there is a week-long holiday called *Maslenitsa* to mark the end of winter.

Chapter 14, Section 2

Life in Russia (continued)

Explaining

What are two challenges that Russia faces?

1. _____
_____

2. _____
_____

Transportation and Communications

Railroads are the major way of moving people and goods throughout the vast land of Russia. A network of railroads covers the heavily populated area west of the Ural Mountains. This network connects to the Trans-Siberian Railroad. Completed in the early 1900s, the Trans-Siberian Railroad is the longest rail line in the world. It runs from Moscow in the west to Vladivostok in the east. The railroad makes it possible for Russians to use Siberia's natural resources.

Russia does not have a good highway system. No multilane highways connect cities. The roads that do exist are in poor condition. Car ownership is on the rise in Russia, so more and better roads are needed. The government is currently building a 6,600-mile national highway across the country.

Russia is also working on improving its communications systems. Upgrades have been made in the telephone system, including installing new phone lines for faster transfer of Internet information. Many rural areas still have poor phone service, however.

Section Wrap-Up

Answer these questions to check your understanding of the entire section.

1. **Describing** What effect does the scarcity and expense of housing have on Russian lifestyles?

_____
_____

2. **Explaining** What is the government of Russia doing to solve some of the country's transportation problems?

_____
_____

On a separate sheet of paper, write an oral tradition that has been passed down in your family, including how long the story has been passed down.

Chapter 15, Section 1 (Pages 408–412)
A Changing Russia

Big Idea

Geographers organize the Earth into regions that share common characteristics. As you read, complete the diagram below. Describe three major effects of the fall of communism on Russia.

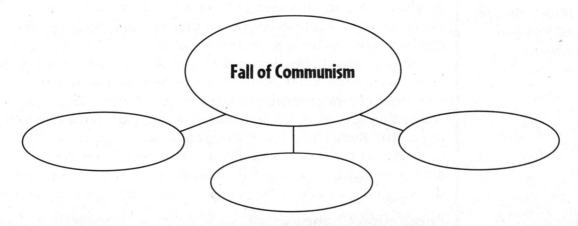

 **Read to Learn**

Changing Politics and Society (pages 409–411)

Specifying
List three aspects of the new Russian government that were determined by Russian voters.

1. _____
2. _____
3. _____

The fall of communism in 1991 brought changes to Russia's government, economy, and society.

A New Form of Government
After communism fell, Russia became more democratic. In 1993 Russians voted on a new constitution, elected representatives to the legislature, and elected their first president—Boris Yeltsin. Today the Russian Federation is a federal republic, with power divided between national and regional governments.

A New Economic System
Russia has tried to move from a command economy to a market economy. One feature of this new economy is privatization. The goal of **privatization** is to shift the ownership of businesses from the government to individuals. Businesses now have to compete with one another to produce goods that Russian consumers need and want—and at a price they are willing to pay.

Changing Politics and Society (continued)

Summarizing

Underline three positive effects of Russia's new political freedom.

Circle three negative effects of Russia's new economic freedom.

Changes in Society

Russians now have political freedom. They may join different political parties and are allowed to criticize leaders and their policies. The government no longer controls news reports and books. Russians have more contact with American and European ideas, music, and fashion. Consumerism, or the desire to buy goods, has led to the emergence of a middle class in Russia. People in the **middle class** are neither rich nor poor. But they can buy cars, electronics, and new clothing.

New economic freedom did not guarantee success, however. Some businesses failed. Some Russians lost their jobs. Other workers face **underemployment,** meaning they work at jobs for which they are overqualified. Many people must have two jobs to survive. Russia also has many **pensioners** who are unable to work and receive a fixed income from the government. In a market economy, prices go up and down. Pensioners' incomes do not go up, so pensioners cannot always afford to buy goods.

Population Changes

During Soviet times, many ethnic Russians moved to other parts of the Soviet Union. When those republics became independent, ethnic Russians were no longer welcome. Many decided to return to Russia. Even so, Russia's population has declined. Low birthrates and rising death rates are the cause.

Russia's Economic Regions (pages 411–412)

Identifying

Identify Russia's four economic regions.

1. _____
2. _____
3. _____
4. _____

The Moscow Region

The city of Moscow is the political, economic, and transportation center of Russia. The Moscow region is home to much of Russia's manufacturing. During Soviet rule, factories focused on **heavy industry,** or producing goods such as machinery, mining equipment, and steel. Today many factories have changed to **light industry,** or producing consumer goods such as clothing and household products.

St. Petersburg and the Baltic Region

St. Petersburg and the Baltic region are located in northwestern Russia, near the Baltic Sea. St. Petersburg is a cultural center, attracting thousands of tourists. The city also is a major port, making it an important trading center. Factories there make ships, machinery, and vehicles. The people of St. Petersburg buy their food and fuel from other regions in Russia.

 | # Read to Learn

Russia's Economic Regions (continued)

Paraphrasing

Complete the following sentences.

The Volga River is in the _____ and _____ region.

The river provides water for _____ and for _____ farms.

Another Russian port city on the Baltic Sea is Kaliningrad. It is located on a small piece of land between Poland and Lithuania, isolated from the rest of Russia. Goods shipped to Kaliningrad from other parts of Russia actually have to cross through other countries to get to this port. Kaliningrad is Russia's only port on the Baltic Sea that stays ice-free all year.

The Volga and Urals Region

To the south and east of Moscow is the Volga and Urals region. This region is valued for its manufacturing and farming. The Volga River is a vital waterway used for shipping people and goods, and it carries nearly half of Russia's river traffic. The Volga River also provides water needed for hydroelectric power and for irrigating farms. Wheat, sugar beets, and other crops are grown in this region. Many of Russia's natural resources come from the Ural Mountains. Minerals such as copper, gold, lead, nickel, and bauxite are found there.

Siberia

Russia's fourth economic region is Siberia. This region holds valuable iron ore, uranium, gold, and coal. Timber is harvested from the taiga. Cold Arctic winds and frozen ground make it difficult for Russia to take advantage of Siberia's resources, however.

Section Wrap-Up

Answer these questions to check your understanding of the entire section.

1. **Comparing and Contrasting** Compare and contrast the Moscow region and the St. Petersburg/Baltic region.

   _____
   _____

2. **Analyzing** What barriers prevent making use of Siberia's natural resources?

   _____
   _____

 *On a separate sheet of paper, write an editorial either for or against Russia's move from a command economy to a market economy. Give reasons to support your opinion.*

Chapter 15, Section 1

Chapter 15, Section 2 (Pages 418–422)
Issues and Challenges

Big Idea

Geography is used to interpret the past, understand the present, and plan for the future. As you read, complete the diagram below. Identify six changes, both positive and negative, that have resulted from Russia's switch to a free market economy.

Read to Learn

Political and Economic Challenges (pages 419–421)

Speculating

What is a decree? In your own words, explain why decrees are undemocratic.

_____
_____
_____
_____
_____

Roadblocks to Democracy

Becoming a democracy has not been easy for Russia's people. Confusion over government powers is a problem. In the new government, the Russian president has the power to issue **decrees,** or rulings that have the force of law but do not need the legislature's approval. This power gives the president much control over the country.

Russia is a federal republic, which means that power is shared among national, regional, and local governments. To ensure that leaders obey his wishes, however, President Vladimir Putin organized the country into seven large districts. He then appointed his supporters as district governors.

Many Russian politicians ignore democratic practices. The courts and legal system tend to favor wealthy citizens. Most of the people do not understand how the government works, which prevents them from being able to change it.

Political and Economic Challenges (continued)

Explaining
Explain why banks are important in an economy.

_____
_____
_____
_____
_____
_____

Identifying
In which region of Russia has a violent separatist movement occurred?

_____

Shifting to a Market Economy

In Russia's market economy, new companies have been started, some incomes have risen, and higher prices for oil and gas exports have brought more money into the country. But oligarchs control various parts of the economy. **Oligarchs** are members of a small group of rulers that holds much power. Russia's oligarchs are often corrupt business leaders.

The benefits of economic success have not reached all the Russian people. Some Russians have become wealthy, but others have become even poorer. Economic success varies widely from region to region, too. Incomes in Moscow are much higher than incomes in other Russian cities.

Russia's banking system has hindered economic growth. Banks play an important role in an economy. People deposit their savings into banks to earn interest. Banks, in turn, loan that money to other people who buy homes or cars, or start new businesses. These actions create jobs. In Russia, however, many people do not trust the banks. They are afraid that they will lose the money they deposit. To remedy this problem, the government created a **deposit insurance** system that promises to repay people who deposit money in a bank if that bank goes out of business.

Challenges to National Unity

Regional rivalries have increased in recent years and make it difficult to unify Russia. When the Soviet Union fell apart, some of the ethnic groups in Russia wanted to form their own countries. This gave rise to **separatist movements,** or efforts to break away from the national government and become independent. A violent separatist movement began in Chechnya, a region near the Caucasus Mountains in southern Russia. Russia's then-President Boris Yeltsin gave the region more self-rule, but many of the Chechen people wanted complete independence. In 1994 the Russian army was sent to Chechnya, and both sides suffered heavy losses. The situation remains unresolved.

Notes | Read to Learn

Russia and the World (page 422)

Finding the Main Idea

What is the main idea of this subsection?

_____
_____
_____
_____

Russia is a major world power and is prominent in world affairs. The country has worked in recent years to help strengthen its connections with other countries. For example, in 2002 Russia supported the United States and other North Atlantic Treaty Organization (NATO) countries in their fight against global terrorism.

At the same time, many countries, including the United States, are troubled by President Vladimir Putin's increased power. They are concerned by Putin's moves away from democratic practices. Tension also is increasing between Russia and former Soviet countries. Some Russian leaders want to have more influence in these countries. Although neighboring countries are unhappy with Russia's actions, they depend on Russia for oil and natural gas.

Section Wrap-Up

Answer these questions to check your understanding of the entire section.

1. **Describing** What challenges face the Russian people as they shift toward democracy?

_____
_____

2. **Explaining** How have regional rivalries hurt Russia?

_____
_____

In the space provided, write a paragraph describing Russia's role as a world power and predicting the role Russia will play in world affairs over the next several years.

_____
_____
_____
_____
_____

Chapter 16, Section 1 (Pages 442–446)
Physical Features

Big Idea

The physical environment affects how people live. As you read, complete the chart below by listing five bodies of water or landforms of the region. Then explain why each is important.

Physical Feature	Importance
1.	
2.	
3.	
4.	
5.	

Notes — Read to Learn

The Region's Landforms (pages 443–444)

Locating

Identify the mountain ranges located in each area below.

North Africa

_____

_____

Southwest Asia

_____

_____

Central Asia

_____

_____

North Africa, Southwest Asia, and Central Asia extend from the Atlantic coast of northern Africa to the mountains in the middle of Asia.

Seas and Waterways

The major bodies of water that surround the region include the Atlantic Ocean, the Mediterranean Sea, the Black Sea, the Red Sea, the Persian Gulf, and the Arabian Sea. These seas have enabled the people to trade more easily with the rest of Africa, Asia, and Europe. Smaller waterways link these seas. They include the Strait of Gibraltar, the Dardanelles Strait, the Sea of Marmara, the Bosporus Strait, the human-made Suez Canal, and the Strait of Hormuz.

Mountains, Plateaus, and Lowlands

The landscape throughout the region is rugged. The Atlas and Ahaggar Mountains cover much of western North Africa. Low plains and low-lying plateaus make up the central and eastern parts of North Africa. The Zagros Mountains and Hindu Kush rise in Southwest Asia. A narrow gap through the Hindu Kush, called the Khyber Pass, is part of a trade route linking

Chapter 16, Section 1   109

The Region's Landforms (continued)

Identifying

Underline two civilizations that arose in this region. Circle the rivers that allowed these civilizations to thrive.

Southwest Asia to other parts of Asia. Central Asia holds the lofty Pamirs and Tian Shan ranges. Central Asia also has lowlands along the Caspian Sea as well as several desert areas.

Rivers

People have long settled along river valleys to benefit from the rich soil. The civilization of Egypt arose along the Nile River. Flooding of the Nile River provided water and **silt,** or small particles of rich soil. This made the land good for growing crops. The Tigris and Euphrates Rivers in Southwest Asia formed an **alluvial plain,** or an area of fertile soil left by river floods. The civilization of Mesopotamia arose on this alluvial plain.

Natural Resources (pages 444–446)

Listing

Write down seven natural resources found in North Africa, Southwest Asia, and Central Asia.

1. _____
2. _____
3. _____
4. _____
5. _____
6. _____
7. _____

The region of North Africa, Southwest Asia, and Central Asia is rich in natural resources. Two of its resources, petroleum and natural gas, are important to countries around the world.

The largest reserves of petroleum (oil) and natural gas are in the Persian Gulf. The land in the area is made up of **sedimentary rock,** a type of rock created when layers of material are hardened by extreme weight. For millions of years, oil collected between the layers of rock.

Countries with oil have become wealthy from selling it. They have used this wealth to develop industry and to provide benefits to their people. Countries in the region without oil have remained poor. As the people in the oil-rich nations interact more with people from other countries, their cultures are exposed to new and different ideas. Conflicts sometimes develop between people who prefer the traditional ways of life and those who support new ways of life.

Other natural resources in this region are coal, iron ore, and **phosphates,** or mineral salts used to make fertilizer. Forests are scarce, except in Lebanon, which has a lumber industry. Fish are a plentiful resource in parts of the region.

Environmental Concerns—The Seas

The misuse of water in North Africa, Southwest Asia, and Central Asia has damaged both the water and the land. Because water is scarce in the region, any misuse will cause problems.

The Caspian Sea has been harmed by overfishing. Illegal fishing, called **poaching,** has decreased the number of sturgeon.

In the 1960s, irrigation projects drained the two rivers that feed the Aral Sea, causing it to start drying up. The water became

Notes | Read to Learn

Natural Resources (continued)

Identifying
Write a sentence identifying three threats to the water supply in the region.

_____
_____
_____
_____

saltier and unfit for drinking. Fish populations also were harmed. Steps are being taken to save the Aral Sea. Dams and dikes have been built to help raise the water level. A higher water level lowers the salt level.

Other Environmental Issues

Another misuse of water occurs when irrigation water evaporates and leaves salt deposits on the land. The salt makes the land less fertile. In severe cases, the land no longer can be used for farming.

Dams built to control flooding have had positive and negative effects. The Aswān High Dam controls the Nile River's floodwaters. This helps Egyptian farmers and also provides hydroelectric power for Egypt's cities and factories. The dam blocks the flow of silt, however, so now the farmers have to use fertilizers to enrich the soil.

Air pollution is a growing problem. Older cars in the region release polluting fumes. Refineries also release chemicals that pollute the air. **Refineries** are facilities that process petroleum to make gasoline and other products.

Section Wrap-Up
Answer these questions to check your understanding of the entire section.

1. **Describing** Describe one positive effect and one negative effect of North Africa, Southwest Asia, and Central Asia being an oil-rich region.

   _____
   _____

2. **Determining Cause and Effect** How has the drying up of the Aral Sea affected human and animal populations? What steps are being taken to correct the problems?

   _____
   _____

Expository Writing *On a separate sheet of paper, write a paragraph describing the significance of seas and inland waterways to this region.*

Chapter 16, Section 1

Chapter 16, Section 2 (Pages 447–450)
Climate Regions

Big Idea

Places reflect the relationship between humans and the physical environment. As you read, complete the web diagram below by identifying five climates in the region.

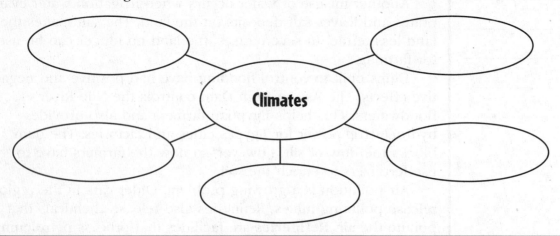

Notes — Read to Learn

A Dry Region (pages 448–449)

As you read, list the four deserts mentioned.

_____

_____

_____

_____

Much of the land in North Africa, Southwest Asia, and Central Asia is desert, with a hot, dry climate. The main cause of this climate is the dry continental air masses that blow over much of the region.

The Sahara

The Sahara is the world's largest desert, and it covers most of North Africa. The Sahara gets only about 3 inches of rain per year. Most of this rain falls in the winter months. Sometimes violent summer thunderstorms cause flooding. **Wadis,** or dry riverbeds, fill with water when it rains.

Most of the Sahara is rock or gravel. About 20 percent of the Sahara is covered by **ergs,** or large sand dunes. Oases also appear in the Sahara. **Oases** are places where the land is fertile as a result of water from a spring or a well. Villages, towns, and cities are built around oases.

Desert and Steppe Areas

Other deserts appear throughout this region. The Rub' al Khali, or Empty Quarter, is a vast desert in the southern part of the Arabian Peninsula. It averages about 4 inches of rain per year.

A Dry Region (continued)

Contrasting

What is the difference between the two major lifestyles of people who live in steppe areas?

_____
_____
_____
_____
_____
_____
_____
_____
_____

Two large deserts in Central Asia are the Kara-Kum and Kyzyl Kum. The deserts in Central Asia are in the middle latitudes. This location causes these deserts to have hot summers but very cold winters.

Dry, treeless, grassy plains called **steppes** border the deserts. Steppes lie north of the Sahara and also can be found in Turkey and to the east in portions of Central Asia. Steppes receive 4 to 16 inches of rain per year.

Some people who live in the steppes are **nomads** who move constantly to find food and water for their herds of animals. Other people are settled and dry farm. **Dry farming** is a method in which land is left unplanted every few years in order to store moisture.

Other Climate Areas

The Mediterranean climate can be found in the coastal areas of North Africa, the eastern Mediterranean, and Turkey. Summers in these areas are hot, but they receive enough rainfall to make the land good for agriculture. As a result, more people live in these areas than in other parts of the region.

A humid subtropical climate covers a small part of Central Asia. Here the summers are hot, the winters are mild, and plenty of rain falls.

The mountainous areas of the region have highland climates. People who live in the highlands tend to herd animals, because the land is difficult to farm.

The Need for Water (page 450)

Summarizing

Summarize three ways that countries try to prevent or deal with their water shortages.

_____
_____
_____

Access to water is a problem in North Africa, Southwest Asia, and Central Asia. The region receives little rainfall, and the high temperatures evaporate surface water quickly. A large amount of water is needed to irrigate farmland. Some countries have turned to aquifers as a water source. **Aquifers** are underground rock layers through which water flows.

Countries often have to compete for water, and this can lead to conflicts. For example, Turkey has built dams on the Tigris and Euphrates Rivers. These dams reroute water to Turkey that otherwise would have flowed to Syria and Iraq.

Some countries ration water as a way to manage the shortage. **Rationing** is a method of making a resource available to people in limited amounts. Other countries remove salt and minerals from seawater through a process called **desalinization**. Wealthy countries have built desalinization plants to provide drinkable

Notes | Read to Learn

The Need for Water (continued)

water for their people and usable water for irrigation. Desalinization is expensive, however, and poor countries cannot afford to do it. They will continue to face water shortages.

Section Wrap-Up

Answer these questions to check your understanding of the entire section.

1. **Defining and Explaining** What are oases? Why would people tend to live around them?

   _____
   _____
   _____

2. **Drawing Conclusions** How can the shortage of water lead to conflict between countries?

   _____
   _____
   _____

Descriptive Writing

In the space provided, write a paragraph describing what you think a nomadic life might be like.

_____
_____
_____
_____
_____
_____
_____
_____

Chapter 17, Section 1 (Pages 456–465)
History and Religion

Big Idea

The characteristics and movement of people affect physical and human systems. As you read, complete the time line below. List four key events in the development of the world's three major religions.

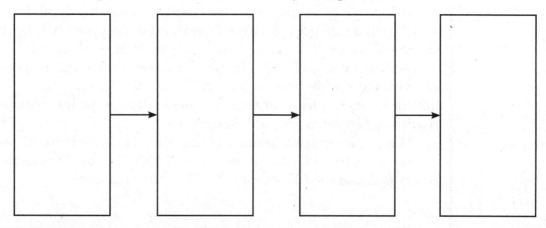

Notes | Read to Learn

Early Civilizations (pages 457–458)

As you read, write the bold terms in the proper spaces in the Venn diagram.

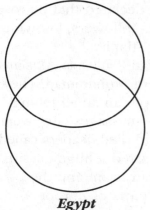

Mesopotamia

Around 4000 B.C., people began settling along the Tigris and Euphrates Rivers—part of a region known as the Fertile Crescent. The settlements became the civilization of Mesopotamia. The land was made fertile by the yearly flooding of the rivers. Farmers used **irrigation** to bring water from the rivers to their fields. Irrigating their crops let Mesopotamians grow a larger, more stable supply of food, and the population expanded.

By 3000 B.C., many cities had formed in the Sumer region of southern Mesopotamia. Each Sumerian city and the land around it became a **city-state** with its own government. Each city-state had a **theocracy,** or a government controlled by religious leaders. Religion was based on **polytheism**—the worship of many gods.

The Sumerians created one of the first calendars and developed an early form of writing called **cuneiform.** They were the first people known to use the wheel and the plow. With few natural barriers, however, powerful rulers easily invaded the region. One such ruler was Hammurabi, who conquered Mesopotamia about 1790 B.C. He developed one of the world's first written codes of law to protect people and their property.

Early Civilizations (continued)

Defining
What was a pharaoh?

_____
_____
_____

Ancient Egypt

Around 5000 B.C., the civilization of Egypt grew along the Nile River in northeastern Africa. Egyptians depended upon the Nile for water and irrigation. The river's yearly floods enriched the soil and led to increased food supplies and a growing population. Egypt was united under a single ruler around 3100 B.C.

Religion was at the center of Egyptian society. Egyptians worshiped many gods and goddesses. Like Mesopotamia, Egypt was a theocracy. The ruler's title was **pharaoh,** and he was believed to be a god. The pharaohs ordered thousands of people to build temples and pyramids, or tombs. The Egyptians also developed a form of writing called **hieroglyphics,** which used pictures for words and sounds.

One of the greatest trading empires of the ancient world was Phoenicia (present-day Lebanon). Around 1000 B.C., the Phoenicians traveled and traded all across the Mediterranean Sea.

Three World Religions (pages 459–461)

Naming
Write the names of the three major religions and their founders or important leaders.

1. _____
   _____
2. _____
   _____
3. _____
   _____

Judaism, Christianity, and Islam began in Southwest Asia. All three are based on **monotheism,** the belief in one God.

According to the Tanakh, or Hebrew Bible, Jews are descended from Abraham, a herder in Mesopotamia. About 1800 B.C., God made a **covenant,** or agreement, with Abraham. If Abraham moved to Canaan, he and his descendants would be blessed. God later revealed His laws, including the Ten Commandments, to Moses, a **prophet,** or messenger of God. About 1000 B.C., King David created a kingdom in Israel.

A Jewish teacher named Jesus began preaching in Israel about A.D. 30. He taught that if people placed their trust in God, their sins would be forgiven. Some viewed Jesus as a savior sent by God. This worried other Jews and Roman rulers, who had Jesus crucified for treason. Jesus' followers declared that he rose from the dead and was the Son of God. His followers, known as Christians, set up churches and spread Christianity.

Islam began in the A.D. 600s in the Arabian Peninsula. Muslims, or members of Islam, follow the teachings of Muhammad. They believe that Muhammad heard messages from an angel telling him to preach about one God, Allah. These messages later were written down in the Quran. After Muhammad died, leaders called **caliphs** ruled the Muslims. Their armies created a huge empire. Muslim merchants and traders helped spread Islam and the Arabic language to Asia, North Africa, and parts of Europe.

The Region in the Modern World (pages 462–465)

Formulating Questions

Write a question about North Africa, Southwest Asia, and Central Asia in the modern world. Then answer your question.

_____
_____
_____
_____
_____

In the late A.D. 900s, the region of North Africa, Southwest Asia, and Central Asia was a vast Arab Empire. But during the next centuries, Mongol invaders swept into the Muslim world from Central Asia. In the late 1200s, a new Muslim empire arose, led by the Ottoman Turks. The Ottoman Empire lasted until its defeat in World War I.

After World War I, large areas of the region came under the control of European powers. The people resented foreign control. Through wars and political struggles, most countries in Southwest Asia and North Africa won independence by the 1970s. Central Asian countries gained independence after the Soviet Union fell in 1991.

In 1947 the United Nations (UN) voted to divide Palestine into Jewish and Arab countries. In 1948 the state of Israel was established. The Arabs considered Palestine their homeland and bitterly opposed the UN decision. Arab-Israeli conflicts have occurred ever since.

Many people in the region have joined political movements that blame America and Europe for keeping Muslim nations poor and weak. A result of this development has been the spread of **terrorism,** or the use of violence against civilians to achieve a political goal.

Section Wrap-Up *Answer these questions to check your understanding of the entire section.*

1. **Drawing Conclusions** Why did the ancient civilizations of Mesopotamia and Egypt develop along rivers?
_____
_____

2. **Contrasting** What is the difference between polytheism and monotheism?
_____
_____

On a separate sheet of paper, write a paragraph explaining how this region changed after World War I.

Chapter 17, Section 1

Chapter 17, Section 2 (Pages 468–476)
Cultures and Lifestyles

Big Idea

Culture groups shape human systems. As you read, complete the diagram below. Summarize the culture of this region by adding one or more facts to each of the ovals.

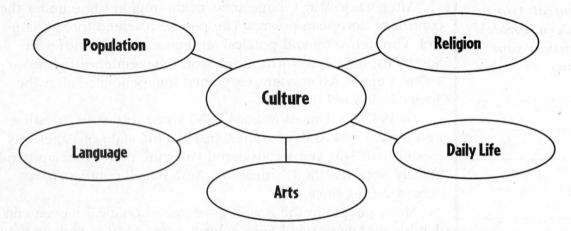

Notes | Read to Learn

Population Changes (page 469)

Describing

Describe four challenges facing cities of this region.

_____

_____

_____

_____

The harsh environment of this region influences where people settle. Most people live near water—along seas and rivers, near oases, or in the rainy highlands. Few people live in the vast deserts, except those who work near oil reserves.

The region's population is growing rapidly because of improved health care for infants and adults, along with a high birthrate. This fast growth is creating challenges. Farming in rural villages is difficult and cannot support the growing population. As a result, many people are moving to the cities, looking for better lives. Large cities, such as Istanbul, Cairo, Tehran, and Baghdad, are important political and economic centers. Yet they face overcrowding, job shortages, lack of transportation, and poor housing. Many people live in poverty on the outskirts of these cities.

Religion, Language, and Arts (pages 471–473)

Listing

What are the Five Pillars of Islam?

1. _____

2. _____

3. _____

4. _____

5. _____

Religion continues to play a significant role in the region. Islam is the major religion. It is divided into two main groups—Sunni and Shia—and they disagree on how Islam should be governed. Most Jews in the region live in Israel. Christians are the dominant group in Armenia and Georgia.

Religious Practices

Each religion has its own practices. Muslims try to fulfill the Five Pillars of Islam. The first pillar, or duty, is a statement of faith. Muslims must pray five times a day while facing Makkah, and they must give money to people in need. They fast during the holy month of Ramadan. The fifth pillar is to undertake a holy journey, or **hajj,** to Makkah to pray.

Christians celebrate Easter and holy days that honor **saints,** or Christian holy people. Orthodox Christians have rituals that go back centuries. Jewish people honor the Sabbath from sundown on Friday to sundown on Saturday. Both Jews and Muslims observe **dietary laws** that state what foods they can eat and how their food must be prepared.

Languages, Literature, and the Arts

As Islam spread, so did Arabic—the major language in the region. Hebrew, Turkish, and Farsi also are spoken. Armenians and Georgians have their own languages. Great works of literature have been written in the region's languages. Many are **epics,** or tales or poems about heroes.

Religion also has influenced the arts in the region. **Mosques,** or Islamic houses of worship, have large interiors, highly decorated surfaces, and brilliant colors. Islam discourages showing human forms in art. Muslim artists instead use geometric patterns, flowers, and beautiful writing known as **calligraphy** in their works. Christian stone churches with domed roofs can be seen in Georgia and Armenia. Carpets are another popular art form. They are handmade with complex designs and rich colors.

Daily Life (pages 474–476)

What percentage of the region's people live in cities?

_____

Rural and Urban Lifestyles

During the last century, many people moved to cities. Today more than half of the region's people live in urban areas. In the largest cities, people live in high-rise apartments. But in older neighborhoods, people still live in stone or mud-brick buildings, some of which lack running water and electricity.

Chapter 17, Section 2   119

Notes | Read to Learn

Daily Life (continued)

Identifying

List three wealthy countries in this region.

_____
_____
_____

List three developing countries in this region.

_____
_____
_____

In rural areas, people grow their own food or shop at the village market. City dwellers can shop in supermarkets, but many people still shop at a **bazaar**, or traditional marketplace.

Living Standards

Countries with an economy based on manufacturing or oil have relatively high standards of living. Israel, for example, has highly skilled workers and exports high-technology products. Saudi Arabia and Qatar are wealthy from oil production. People in oil-rich countries live in modern cities and receive free education and health care from the government.

In the region's developing countries, there is little wealth to share. High population growth has strained the economies of some countries, such as Egypt and Algeria. Many North Africans have migrated to Europe to find jobs. People in Afghanistan and Tajikistan still farm or herd to make a living.

Family Life and Education

Family life is valued in this region, with families gathering at midday for their main meal. Men have the dominant family role. Women are expected to obey their husbands, stay home, raise children, and dress modestly. Women in Saudi Arabia cannot vote, drive, or travel unless a male relative is with them. In other countries, however, women have improved their status.

Section Wrap-Up

Answer these questions to check your understanding of the entire section.

1. **Determining Cause and Effect** Why is the region's population growing rapidly?

   _____
   _____

2. **Analyzing** Why do some countries in the region have relatively high standards of living while others do not?

   _____
   _____

On a separate sheet of paper, write a paragraph explaining how lifestyles vary throughout the region.

Chapter 18, Section 1 (Pages 484–488)
North Africa

Big Idea

Changes occur in the use and importance of natural resources. As you read, complete the diagram below. List the major agricultural and industrial products.

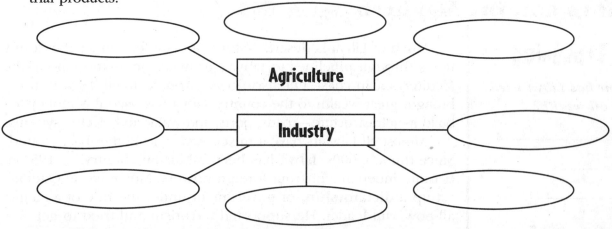

 **Read to Learn**

Egypt (pages 485–486)

Stating

What are two energy resources in Egypt, and where do they come from?

1. _____
_____

2. _____
_____

Large deserts cover Egypt, so most Egyptians live within 20 miles of the Nile River. Mud-brick villages, ancient ruins, and modern cities line the Nile's banks. Some 11 million people live in Cairo, the capital.

Egypt has a developing economy. About one-third of the people farm. Some farmers use modern machinery, but peasant farmers called **fellahin** use simple tools. The best farmland lies in the Nile Valley. The Aswān High Dam controls the yearly flooding of the Nile. Water stored behind the dam is released several times a year. This allows farmers to harvest more than one crop per year. The dam also provides Egypt with hydroelectric power.

Egypt's main energy resource is oil found in and around the Red Sea. Petroleum and **phosphates,** or minerals used in fertilizers, make up Egypt's major exports. Workers also make food products, textiles, and other consumer goods. Industry is drawing people to cities. Poverty, traffic, and pollution plague Cairo and the port of Alexandria.

Europeans and Egyptians worked together to build the Suez Canal in the 1800s. The canal is one of the world's most important waterways, and in time it came under control of the British.

Notes | Read to Learn

Egypt (continued)

In 1952 the British-supported king of Egypt was overthrown, and Egypt became independent. Today Egypt is a republic. Most of the people are Muslims and speak Arabic.

Libya and the Maghreb (pages 486–488)

Explaining

How has Libya used its oil wealth?

_____
_____
_____

Comparing

What three things do the Maghreb people have in common?

1. _____
_____
2. _____
_____
3. _____

Locating

Where are Tunisia's, Algeria's, and Morocco's populations concentrated, and why?

_____
_____
_____

Much of Libya is desert. Water from aquifers under the desert flows through pipelines to Libya's growing population along the Mediterranean coast. Oil also flows through pipelines and has brought great wealth to the country. Libya has used this money to build its infrastructure—roads, ports, and water and electric systems.

Almost all Libyans have a mixed Arab and Berber background. Since the A.D. 600s, Libya has been a Muslim country. In 1951 it became independent from foreign powers. Muammar al-Qaddafi set up a **dictatorship,** or government under the rule of a single, all-powerful leader. He supported terrorism and tried to get nuclear weapons. The United States and the United Nations used **trade sanctions,** or trade barriers, to punish Libya. Qaddafi was forced to change his policy.

The Maghreb

North Africa also includes the Maghreb countries—Tunisia, Algeria, and Morocco. *Maghreb* is Arabic for "the land farthest west." These countries are the westernmost part of the Arabic-speaking Muslim world. All three have mixed Arab and Berber populations.

North Africa's smallest country is Tunisia. Farmers along the fertile coast grow wheat, olives, fruits, and vegetables. Factories produce food, textiles, and petroleum products. Tunisia was part of several Muslim empires. It became a colony of France until gaining independence in 1956. Since then, it has been a republic governed by powerful presidents.

North Africa's largest country is Algeria. Farmers along its Mediterranean coast grow wheat, barley, olives, oats, and grapes. The capital, Algiers, is known for its **casbahs,** or older sections with narrow streets and bazaars. Algeria's economy is based on oil and natural gas pumped from the Sahara. Still, many Algerians have moved to Europe to find work.

In 1830 Algeria became a French colony. A **civil war,** or conflict between groups within a country, erupted in 1954. Algeria won independence when the fighting ended in 1962. It is now a republic with a president and a legislature. Another civil war erupted in the 1990s. The government has tried to restore order since the fighting ended in 1999.

Notes | Read to Learn

Libya and the Maghreb (continued)

Morocco lies in Africa's northwest corner. Farmers there raise livestock and grow grains, vegetables, and fruits on the coastal plains. Morocco also produces phosphates. In addition, tourism has become important. Visitors come to see the historic cities of Marrakesh and Casablanca. Morocco has a **constitutional monarchy**—a king is head of state, but elected officials run the government. In 1975 Morocco seized Western Sahara, a mineral-rich desert region to the south. Western Saharans have fought for independence from Morocco.

Section Wrap-Up

Answer these questions to check your understanding of the entire section.

1. **Listing** What are the five countries of North Africa, and what type of government does each one have?

   _____

   _____

2. **Explaining** What are trade sanctions? Why were they used against Libya?

   _____

   _____

On the lines below, write a paragraph describing an imaginary trip to a casbah in Algeria. What do you see, hear, and smell?

_____

_____

_____

_____

_____

_____

Chapter 18, Section 1

Chapter 18, Section 2 (Pages 490–498)
Southwest Asia

Big Idea

Cooperation and conflict among people have an impact on the Earth's surface. As you read, complete the diagram below. List key facts about the economies of Southwest Asia.

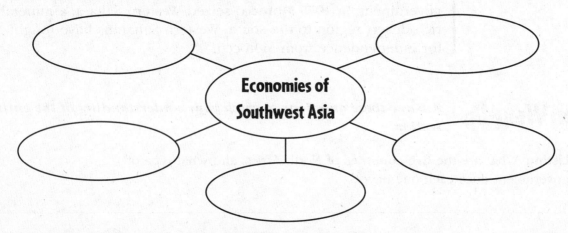

Notes | Read to Learn

The Eastern Mediterranean (pages 491–494)

Listing

What are some crops grown in the eastern Mediterranean region?

_____
_____
_____
_____

Identifying

Underline the cultural influences found in Lebanon.

Turkey bridges the continents of Asia and Europe. Its Mediterranean climate allows farmers to grow enough food for the country and to export cotton and tobacco. Textiles, steel, and cars also support the economy. Most Turks are Muslim, but many prefer a **secular,** or nonreligious, society. The largest city, Istanbul, is filled with palaces and mosques. Once the heart of the Ottoman Empire, Turkey today is a republic.

Syria, south of Turkey, has mountains, deserts, and coastal plains. Most Syrians live in rural areas and grow cotton, wheat, and fruit. The Euphrates River supplies water for irrigation and hydroelectricity for cities and industries. Damascus, the capital, was founded more than 4,000 years ago. Syria's one-party government restricts political freedoms.

Small Lebanon's fertile Mediterranean coast allows farming. However, most Lebanese work in or near the capital, Beirut, which has banking, insurance, and tourism industries. Lebanese culture blends Arab, Turkish, and French influences. In 2006 Muslims in Lebanon clashed with Israel.

In western Jordan, farmers grow wheat, fruits, and vegetables in the Jordan River valley. People there also work in service and

The Eastern Mediterranean (continued)

Labeling

Write moshav *and* kibbutz *in the proper places below.*

Shared Property & Work:

1. _____

Private Property, Shared Work:

2. _____

manufacturing industries in cities such as Amman. Jordan's eastern deserts are home to **bedouins**—nomads who live in tents and raise livestock. Jordan's rulers, such as King Hussein I, have tried to blend traditional and modern ways.

Israel became a Jewish republic in 1948. It has a developed economy that produces high-technology equipment, clothing, chemicals, and machinery. Advanced irrigation systems help farmers grow citrus fruits, vegetables, and cotton. Some Israeli farmers live on a **kibbutz**, where they share the work and property, and some live on a **moshav**, where they share the work but own some private property.

In 1993 Israel gave Palestinians limited self-rule in the Gaza Strip and parts of the West Bank. Some Muslim Palestinians do not recognize Israel's right to exist and have bombed Israel.

The Arabian Peninsula (pages 494–495)

Stating

What three landforms are found in Saudi Arabia?

1. _____
2. _____
3. _____

Most of Saudi Arabia is covered by deserts. Highlands with fertile valleys lie in the south. Most Saudis live along the Red Sea and Persian Gulf or near desert oases. Skyscrapers rise in many cities, including the capital, Riyadh, which sits amid a large oasis. Saudi Arabia is one of the world's leading oil producers. Oil wealth has been used to build schools, hospitals, and roads, and to improve the standard of living. Saudi Arabia has a monarchy that in 1932 united the country's **clans**, or groups of families related by blood or marriage. The government prepares Makkah and Madinah for the several million Muslims who visit each year.

Kuwait, Bahrain, Qatar, and the United Arab Emirates are located on the Persian Gulf. They export oil and use the money to build their economies and provide free education, health care, and other services. In addition, Qatar has a natural gas industry, and Bahrain is a banking center. Dubai is a large port, financial center, and tourist resort in the United Arab Emirates. Bahrain, Kuwait, and Qatar are moving toward democracy.

Oman is a desert country on the Arabian Peninsula. It exports oil, and has used the resulting wealth to build ports for oil tankers and to build up a tourism industry. Yemen, also on the peninsula, has little oil. Most of its people either farm or herd sheep and cattle. Aden is Yemen's major port city.

Chapter 18, Section 2

Iraq, Iran, and Afghanistan (pages 496–498)

Defining

What is an Islamic republic?

_____

_____

_____

The Tigris and Euphrates Rivers flow through Iraq. An **alluvial plain,** or rich soil left by river floods, lies between the rivers. Farmers there grow wheat, barley, rice, vegetables, dates, and cotton. Iraq's factories produce food, textiles, chemicals, and building materials. Iraq also exports oil.

During the last half of the 1900s, Iraq was governed by dictators, including Saddam Hussein. After his invasion of Kuwait in 1991, the United Nations put an **embargo** on Iraq to restrict its trade with other countries. The embargo damaged Iraq's economy. In 2003 American and British forces overthrew Saddam. Violence arose among Iraq's three largest Muslim groups—Shias, Sunnis, and Kurds. The country remains unstable.

Iran is an Islamic republic run by Muslim religious leaders. Most Iranians are not Arab, but are Persians or Azeri. Many Western customs—seen as a threat to Islam—are forbidden. Iran has large oil reserves. Other industries include textiles and metal goods. Farmers grow wheat, rice, sugar beets, nuts, and cotton. Iran's nuclear activities worry Western countries.

Afghanistan is a mountainous, landlocked country east of Iran. The Khyber Pass is a major trade route through the Hindu Kush. Kabul, the capital, lies in a valley. Two of the largest ethnic groups are the Pashtuns and the Tajiks. Most people herd livestock or grow wheat, fruit, and nuts. Wool and handwoven carpets are exported. Local leaders and Taliban militants have kept the new, democratic government from stabilizing the country.

Section Wrap-Up *Answer these questions to check your understanding of the entire section.*

1. **Listing** How do Saudi Arabia, Kuwait, and Oman spend oil money?

_____

_____

2. **Determining Cause and Effect** Why does Iraq have fertile farmland?

_____

_____

On a separate sheet of paper, write a paragraph explaining what an embargo is and how it affected Iraq.

Chapter 18, Section 3 (Pages 504–508)
Central Asia

Big Idea

Places reflect the relationship between humans and the physical environment. As you read, complete the diagram below. Write three facts about this region in the smaller boxes. In the large box, write a generalization from those facts.

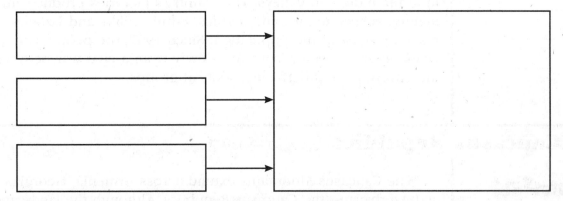

Read to Learn

The Central Asian Republics (pages 505–506)

Listing

What natural resources are found in these countries?

Kazakhstan

_____

_____

Turkmenistan

_____

_____

Kyrgyzstan

_____

_____

The Central Asian Republics are Kazakhstan, Uzbekistan, Turkmenistan, Kyrgyzstan, and Tajikistan. All are Islamic countries and were part of the Soviet Union until its collapse in 1991.

Much of Kazakhstan is covered by dry, treeless plains that are good for raising livestock. The country is rich in copper and petroleum. Half the people are ethnic Kazakhs. They were horse-riding nomads until the Soviets forced them to live in settled areas. Russians are the second-largest ethnic group. Kazakhstan is not a democracy, but it has a free market economy. Foreign investment has boosted the economy.

Many of Uzbekistan's people live in fertile valleys and oases. The economy relies on agriculture, particularly cotton. It is the major **cash crop,** or farm product grown for export. Irrigating cotton fields drained away the rivers flowing into the Aral Sea, reducing its size. Tashkent, the capital, is Central Asia's largest city and industrial center.

Most of Turkmenistan is covered by the Kara-Kum desert. Many Turkmen live in oases and grow cotton or raise livestock. Turkmenistan has large petroleum and natural gas deposits. A powerful president rules from Ashkhabad, the capital.

Chapter 18, Section 3   **127**

The Central Asian Republics (continued)

Summarizing

Summarize the stability of the Central Asian Republics.

_____
_____
_____

Kyrgyzstan is mountainous. Farmers there grow cotton, vegetables, and fruits in valleys and plains. The country has mercury and gold but little industry. More than half the people are ethnic Kyrgyz, but Russians, Uzbeks, and Ukrainians also live there. A revolt overthrew the government in 2005. New leaders have promised democratic reforms.

Tajikistan's farmers raise cotton, grapes, grain, and vegetables in fertile mountain valleys. The country's factories produce aluminum, vegetable oils, and textiles. Ethnic Tajiks and Uzbeks make up most of the population. Since 1997, the people and economy have been recovering slowly from a civil war between the government and Muslim political groups.

The Caucasus Republics (pages 507–508)

Expressing

What three features do ethnic Georgians take pride in?

_____
_____
_____

Specifying

What religions are practiced in Armenia and Azerbaijan?

Armenia

_____

Azerbaijan

_____

The Caucasus Mountains extend across Armenia, Georgia, and Azerbaijan—the Caucasus Republics. Although the landscape is mountainous, the climate is generally mild. Farmers grow tea, citrus fruits, wine grapes, and vegetables in river valleys. All three countries were part of the Soviet Union until 1991.

Georgia borders the Black Sea. It has deposits of copper, coal, manganese, and oil in its mountains. Hydroelectricity powers the country's vehicle, steel, cotton, and textile industries. Georgia's capital, Tbilisi, is located in an area where tectonic plates collide. As a result, it has mineral springs. These hot springs, as well as resorts along the Black Sea coast, make Georgia popular with tourists. Most people are ethnic Georgians. They take pride in their language, culture, and Christian background. Other ethnic groups in the country want independence, however, which has led to conflict.

Armenia is a landlocked country south of Georgia. It sits on many **faults,** or cracks in the Earth's crust. Thus, Armenia experiences frequent, serious earthquakes. The people are mostly ethnic Armenians, and they share a unique language and ancient culture. The official religion is Christianity. Throughout its history, Armenia has been ruled by other people. During World War I, the Ottoman Turks killed many Armenians in a **genocide,** or the deliberate killing of an ethnic group. Some ethnic Armenians live in an **enclave**—a small territory surrounded by a larger territory—in Azerbaijan. Armenia sent an army to protect these people in a dispute that continues today.

The Caucasus Republics (continued)

The country of Azerbaijan and its capital, Baku, border the Caspian Sea. The majority of the people are Azeris who practice Shia Islam. The country has a developing economy. Farmers grow grains, cotton, and grapes for wine. Oil and natural gas deposits under the Caspian Sea will benefit the economy. Azerbaijan is working with companies from other countries to make use of these natural resources.

Section Wrap-Up *Answer these questions to check your understanding of the entire section.*

1. **Identifying** What are two characteristics shared by the Central Asian Republics?

   _____
   _____

2. **Explaining** How have ethnic groups affected the Caucasus Republics?

   _____
   _____

In the space provided, write a paragraph explaining why you think the countries of Central Asia and the Caucasus Republics play an important role in today's world.

_____
_____
_____
_____
_____
_____
_____

Chapter 18, Section 3

Chapter 19, Section 1 (Pages 530–535)
Physical Features

Big Idea

Physical processes shape Earth's surface. As you read, complete the web diagram below. List the physical forces that have shaped the landforms of Africa.

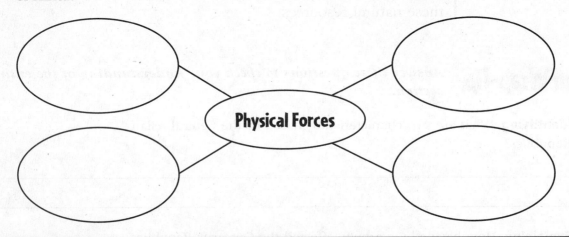

Notes | Read to Learn

Landforms of Africa South of the Sahara (pages 531–532)

Name two specific landforms found in eastern Africa and one found in southern Africa.

Eastern Africa:

_____

_____

Southern Africa:

_____

Africa south of the Sahara is made up of four subregions—West Africa, Central Africa, East Africa, and Southern Africa. It also includes many islands, most of which were formed by volcanoes.

A series of plateaus rise like steps from west to east across the continent. In eastern and southern Africa, steep, jagged cliffs called **escarpments** mark the edges of plateaus. Rivers that flow across plateaus drop suddenly at escarpments, making it impossible for ships to travel between the interior of the continent and the sea.

Lowland plains border the Atlantic Ocean and Indian Ocean coastlines. Among the plateaus are other sunken areas called basins, which were formed when tectonic activity lifted up the land around them. The Congo Basin is Africa's largest lowland area.

Africa south of the Sahara has a high elevation, but only a few long mountain ranges dominate the landscape. In the east rise the Ethiopian Highlands, along with the volcanic mountain peaks of Kilimanjaro and Mount Kenya. *Kilimanjaro* means "shining mountain" in Swahili. The Drakensberg Range is located in southern Africa.

Notes | Read to Learn

Landforms of Africa South of the Sahara (continued)

The Great Rift Valley cuts through eastern Africa for about 4,000 miles. A **rift valley** is a large break in the Earth's crust formed by shifting tectonic plates. Volcanic eruptions and earthquakes have helped create the Great Rift Valley's spectacular landscape of deep lakes, valley walls, and jagged mountains.

Waterways of the Region (pages 533–534)

Describing

How is the Nile River formed?

_____
_____
_____
_____
_____

Many lakes and rivers provide people with freshwater, fish, and local transportation routes. Most of the lakes in Africa south of the Sahara lie in or near the Great Rift Valley. Lake Tanganyika—at 420 miles long—is the world's longest freshwater lake. However, Lake Victoria is the largest freshwater lake in Africa, and the second-largest in the world after Lake Superior.

Some of the lakes are the sources of rivers. The White Nile originates in Lake Victoria, and the Blue Nile begins at Lake Tana. The White Nile and Blue Nile connect in Sudan to form the Nile—the world's longest river. Three other major rivers are the Congo, the Niger, and the Zambezi. All four rivers begin in the interior and flow toward the sea.

Tectonic formations affect the rivers in this region. Rivers plunge over cliffs and escarpments, creating waterfalls and rapids that make transportation difficult. Victoria Falls plunges 420 feet on the Zambezi River. In turn, Africa's rivers also shape the land. As they flow over the plateaus, they carve out steep-sided valleys called **gorges.**

Mineral Resources (pages 534–535)

Locating

Where are most petroleum and natural gas deposits found in this region?

_____
_____
_____
_____

Plentiful energy resources are found throughout Africa south of the Sahara. Petroleum deposits are mined along the Atlantic coast from Nigeria to Angola, and also in landlocked Chad and Sudan. In many countries, oil has replaced farm products as the major export.

Some countries along the Atlantic also have natural gas deposits. Coal is mined in Nigeria, the Democratic Republic of the Congo, and the Republic of South Africa. In addition, the region's fast-flowing rivers provide hydroelectric power. Akosombo Dam in Ghana formed Lake Volta, one of the world's largest human-made lakes. The dam also supplies hydroelectricity to several West African countries.

Chapter 19, Section 1

Notes | Read to Learn

Mineral Resources (continued)

Explaining

What are two uses of diamonds?

_____

_____

Important mineral resources of this region include metals. Iron ore, chromium, uranium, and copper are abundant. Precious materials also are mined here. The Republic of South Africa is rich in gold and platinum. The Transvaal, a grassy plateau in South Africa, holds a gold deposit more than 300 miles long.

Many gemstones are mined in Africa south of the Sahara. These include diamonds, rubies, emeralds, and sapphires. South Africa is a major diamond producer. Some of the diamonds are used to make jewelry. Others become **industrial diamonds**, which are used to make drills, saws, and grinding tools.

Section Wrap-Up

Answer these questions to check your understanding of the entire section.

1. **Drawing Conclusions** What impact do you think escarpments have had on trade in Africa south of the Sahara?

   _____

   _____

2. **Classifying** Complete this chart by providing examples of the various types of resources found in Africa south of the Sahara.

Type of Resource	Examples
Energy	
Metals	
Precious Materials	

Descriptive Writing

In the space provided, write a paragraph explaining the difference between a basin and a rift valley.

_____

_____

_____

_____

Chapter 19, Section 2 (Pages 538–542)
Climate Regions

Big Idea

Geographers organize the Earth into regions that share common characteristics. As you read, complete the diagram below. In the small boxes, write three facts about the region's climate. In the large box, write a generalization from those facts.

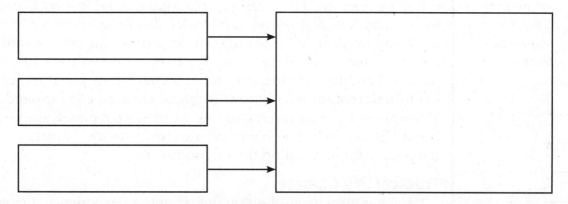

Notes | Read to Learn

Factors Affecting Climate (page 539)

What is the difference in rainfall between the rain forests of Central and West Africa and the Namib Desert?

Rain forest: _____

Namib Desert: _____

Difference = _____

The majority of Africa south of the Sahara lies in the Tropics. Most of the region has high temperatures because it receives the direct rays of the sun year-round. Elevation has an impact on the climate, however. Even in this hot region, mountains have cooler temperatures than the lowland areas.

The climate zones—wet, dry, and temperate—vary throughout the region. The amount of rainfall also varies. Rain forests in Central and West Africa, for example, receive more than 80 inches of rain per year. The Namib Desert in southern Africa gets less than 10 inches of rain per year.

Some areas of Africa south of the Sahara experience long droughts. A **drought** is a period of time when no rain falls at all. Crops fail, and people can starve during severe droughts.

Tropical and Dry Climates (pages 540–541)

Displaying

Note the placement of the word Equator below. As you read, write in the tropical wet, tropical dry, steppe, and desert climate zones in their appropriate places relative to the Equator.

_____
_____
_____
_____

■ **Equator** ■

_____
_____
_____
_____

Where do Africa's most recognizable animals live?

_____

Tropical Wet Climate

The area along the Equator in Central Africa and West Africa has a tropical wet climate. The hot temperatures and plentiful rainfall are ideal conditions for **rain forests**—dense stands of trees and other plants that receive large amounts of rain every year. The floor of a rain forest has mosses, ferns, and shrubs. High above, the tops of the trees form an umbrella-like covering called the **canopy**. Monkeys, parrots, snakes, and insects live in the canopy. Tropical flowers and fruits also grow there.

Many tropical African countries depend on the sale of rain forest products, including wood. Farmers cut down trees for firewood and to clear the land for farming. These practices have led to **deforestation**, or the widespread clearing of forestland. To preserve the rain forests and boost their economies, some countries have developed an **ecotourism** industry. Tourists visit the area without harming the environment.

Tropical Dry Climate

Farther away from the Equator, countries experience a tropical dry climate. In this zone, temperatures are hot, but much less rain falls than in the tropical wet areas. Huge stretches of **savanna**, or grasslands with scattered woods, grow in this dry climate. Some of Africa's most recognizable animals live in these savannas, including elephants, lions, and giraffes. Several countries have set aside their savannas as national parks.

Steppe

Even farther from the Equator is the steppe climate zone. Very little rain falls there, and only for a few months of the year. Plant life includes a variety of trees, thick shrubs, and grasses. Steppe areas are facing **desertificaton,** a process that turns fertile land into land that is too dry to support life. Climate changes, clearing the land for farming, and herding large numbers of livestock have damaged and dried out the land.

Desert

The driest climate zone is desert. Africa has three large deserts—the Sahara in the north and the Namib and Kalahari in the south. The Sahara is mostly rock or stony plains. The Kalahari has vast stretches of sand. The Namib, along the Atlantic coast, is slightly cooler than other deserts because it receives ocean breezes. It also gets moisture from fog. This fog helps support **succulents,** or plants such as cacti with thick, fleshy leaves that conserve moisture.

 **Read to Learn**

Moderate Climate Regions (page 542)

Naming

What three moderate climate zones are found in this region?

1. _____
2. _____
3. _____

Coastal southern Africa and the highlands of East Africa have moderate climates. The temperatures in these areas are comfortable, and there is enough rainfall for farming.

Southeastern Africa has a humid subtropical climate. The summers are hot and rainy, and the winters are mild and rainy. Temperatures become cooler farther south, because the area is farther away from the Equator.

Southwestern Africa has a Mediterranean climate with warm, dry summers and mild, wet winters. Seasons occur at the opposite time of the year from seasons in the United States. Thus, most rain falls during the winter months of June through August. Areas of East Africa with higher elevations have highland climates. Temperatures are cooler, and snow often falls at high altitudes.

Section Wrap-Up *Answer these questions to check your understanding of the entire section.*

1. **Determining Cause and Effect** What practices have led to desertification?

   _____

   _____

2. **Identifying** In the chart below, identify the months of the year for each season south of the Equator.

Seasons	Months

 *On a separate sheet of paper, write a paragraph describing a journey through a rain forest.*

Chapter 19, Section 2   135

Chapter 20, Section 1 (Pages 548–555)
History and Governments

Big Idea

The characteristics and movement of people impact physical and human systems. As you read, complete the diagram below. List features of the early kingdoms and empires that developed in Africa south of the Sahara.

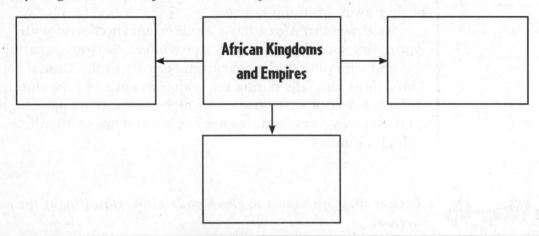

Notes | Read to Learn

Early African History (pages 549–550)

As you read, fill in the boxes with the names of early African peoples and kingdoms.

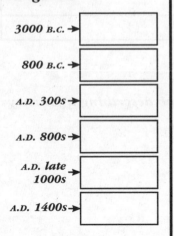

Early Africans were **hunter-gatherers.** They moved from place to place to hunt and gather food. Over time, they became herders and farmers. About 3000 B.C., the Bantu people began migrating from Nigeria to the west and south. They spread their farming and iron-working skills, as well as their language. Today millions of Africans speak Bantu languages.

East and Southern Africa

Kush developed along the Nile River in Sudan around 800 B.C. The people of Kush traded gold, ivory, and iron products. Kush's rulers used the wealth to build temples and monuments like those of Egypt.

The neighboring kingdom of Axum defeated Kush in the A.D. 300s. Located in Ethiopia on the Red Sea, Axum also became wealthy from trade. King Ezana accepted Christianity. In the A.D. 600s, Arab Muslims gained control of the region. Axum remained a center of Christianity.

Farther south, people in the coastal cities of Kilwa, Mombasa, and Mogadishu traded as far as Arabia, India, and China. Arab and African ways blended into the Swahili culture.

Early African History (continued)

Identifying

Who was Mali's most famous ruler?

_____

The empire of Great Zimbabwe arose inland in southeastern Africa. Zimbabwe supplied gold, silver, and ivory to the East African coast during the 1400s.

West Africa's Trading Empires

Three trading empires emerged in West Africa from the A.D. 800s to the 1500s. Ghana controlled and taxed the gold and salt trade between the Sahara and West Africa. This empire fell in the late A.D. 1000s and was replaced by the empire of Mali. Like Ghana, Mali also became wealthy by controlling trade. Mali's most famous ruler, Mansa Musa, made Timbuktu a center of trade, education, and Islamic culture. The kingdom of Songhai took over Mali in the 1400s and lasted until about 1600.

European Contact (pages 551–552)

Determining Cause and Effect

What effects did the European slave trade have on Africa?

_____
_____
_____
_____
_____

The region changed dramatically after Europeans began trading with Africans in the 1400s and 1500s. Slavery already existed in Africa, but it increased greatly when Europeans came. African traders sold captives to Europeans, who shipped the captives to the Americas to be sold into slavery. Between the 1500s and 1800s, nearly 12 million Africans were sent to the Americas. Families were torn apart, villages disappeared, economies collapsed, and kingdoms were weakened.

By the 1800s, Britain declared the slave trade illegal, and other countries followed. Europeans continued to control Africa, however. Businesses wanted Africa's gold, timber, hides, and palm oil. Military leaders and missionaries also had an interest in the region.

In the 1880s, European countries established colonies in Africa south of the Sahara. They carved up the area, tearing apart unified regions and combining ethnic groups that had nothing in common. By 1914, only Ethiopia and Liberia remained independent.

Europeans built railroads and roads and improved farming. They provided some Africans with education and medical care. Europeans did not build many industries, though, because most raw materials were exported. Africans were given fewer rights than Europeans. Many were forced to work in mines or on large farms called **plantations**.

Chapter 20, Section 1

Independence (pages 552–555)

Defining

Explain the words nationalism *and* discrimination *by using them in a sentence.*

_____

_____

_____

_____

Feelings of nationalism arose among European-educated Africans in the 1900s. **Nationalism** is a people's desire to rule themselves and have their own independent country.

After World War I, Africans protested against **discrimination**, the unfair and unequal treatment of a group. European governments eventually made some reforms, but the people of Africa wanted complete independence.

The economic and military demands of World War II weakened European countries. They no longer had the resources to control their colonies in Africa. Africans increased their protests. Kwame Nkrumah led the first successful nationalist movement—Ghana became independent in 1957. By the late 1960s, most African colonies had thrown off European rule.

After achieving independence, many of the new countries kept their old colonial borders. This led to ethnic conflicts and civil wars. Many people became **refugees**, fleeing to other countries to escape mistreatment or disaster. The white-run governments in some countries continued to deny basic rights to non-European people. In South Africa, whites set up a policy called **apartheid**, or "apartness." Laws separated ethnic groups and limited the rights of black South Africans. Many black leaders, including Nelson Mandela, were jailed. Other countries cut off trade with South Africa. Apartheid was ended in the early 1990s.

Section Wrap-Up

Answer these questions to check your understanding of the entire section.

1. **Describing** What was the significance of the Bantu migrations?

_____

_____

2. **Explaining** Why did Europeans give up their African colonies after World War II?

_____

_____

On a separate sheet of paper, write a paragraph explaining how European colonial rule led to the conflicts that occurred after African countries gained their independence.

Chapter 20, Section 2 (Pages 557–564)
Cultures and Lifestyles

Big Idea

Culture groups shape human systems. As you read, complete the diagram below. List information about African society and culture.

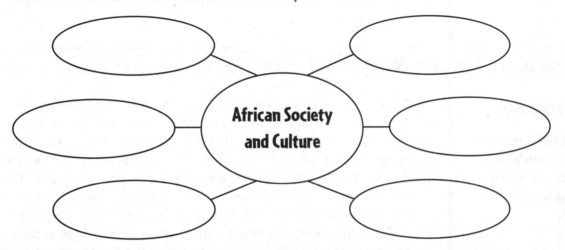

Notes | Read to Learn

The People of Africa South of the Sahara (pages 558–560)

Underline two sentences that tell why a shortage of food exists in the region. Circle the sentence that is addressing the problem.

What does poor sanitation lead to?

_____
_____

The population of Africa south of the Sahara is growing fast. Better sanitation and health care have lowered death rates for infants and children. The region also has a high birthrate. Families average five to seven children. Fast population growth has led to overcrowding. Governments cannot provide enough shelter, water, and electricity. In addition, overuse of the land, soil erosion, and droughts have ruined large areas of farmland. As a result, farmers are unable to grow as much as they once could, and shortages of food exist. Governments are teaching better ways to farm and have tried to convince people to have smaller families.

The population of the region is not evenly distributed. Desert and steppe areas are too dry to support farming or raising livestock. Most people are crowded into coastal areas of West Africa, along the lakes region of East Africa, and along the eastern coast of southern Africa. Cities are growing rapidly as people move from rural areas to find better jobs, health care, and education.

Although health care has improved, the death rate is high compared to other regions. People suffer from **malnutrition,** or poor health due to not eating enough food or the right foods. Some Africans lack clean drinking water and good **sanitation,**

Chapter 20, Section 2   139

The People of Africa South of the Sahara (continued)

or the removal of waste products. Diseases such as malaria are widespread. Many people are infected with the virus that causes AIDS. **Life expectancy,** or the average number of years people expect to live, has gone down—drastically in some countries. Millions of children have been orphaned by AIDS.

Culture in Africa South of the Sahara (pages 561–562)

Describing

Describe the connection between a person's ethnic group and language.

_____
_____
_____
_____

Africa south of the Sahara has a diverse population with many different ethnic groups, languages, and religions. Many Africans feel a strong loyalty to their ethnic group. A person's ethnic group is commonly defined by the language the person speaks. Between 2,000 and 3,000 languages are spoken in the region. However, most are used by a relatively small number of people. Only about a dozen African languages have more than one million speakers. The most widely spoken language is Swahili.

Most people are Christian or Muslim. Yet hundreds of traditional African religions hold beliefs in a supreme being, lesser gods, and the spirits of dead ancestors. Followers of the different religions usually live together peacefully, but conflict between religious groups has occurred recently in Nigeria and Sudan.

African art has many forms and uses. Art often holds religious meaning, tells stories, or depicts the faces of people. Artists make masks and statues out of wood, ivory, and bronze. Weavers design brightly colored cloth. Music and dance also are important. The roles people have in dances indicate their **social status,** or position in the community. Young men and women often dance in ceremonies called **rites of passage,** which mark different stages of life. In West Africa, storytellers called **griots** preserve a group's history by passing down its stories. Written literature has become popular through the works of African authors Chinua Achebe, Wole Soyinka, and Nadine Gordimer.

Daily Life in Africa South of the Sahara (pages 563–564)

Explaining

What occurs at local markets?

_____

About 70 percent of Africans live in rural areas, where they herd livestock or farm. Most farmers grow just enough food for their own families. Any extra food is sold at a local market or traded for other goods. Some farmers work on large farms that grow cash crops for export, such as coffee, cacao, cotton, and bananas.

Daily Life in Africa South of the Sahara (continued)

Contrasting

What is the difference between an extended family and a nuclear family?

_____
_____
_____
_____
_____

Rural families live in villages near their farms. In the past, these families lived in a **compound,** or group of houses surrounded by a wall. Compounds have become less common. Young men often leave their farm villages and work as laborers in the cities.

Cities are centers of industry and commerce. City dwellers have better jobs and, thus, a higher standard of living than rural people. Most people live in single-story homes. The wealthy have luxury apartments or large houses. Many residents live in homes built of wood or concrete blocks on the edge of cities.

Family ties are important in the region. Rural people live in **extended families,** or households made up of several generations. **Nuclear families**—a husband, a wife, and their children—are more common in cities. In addition, African families are organized into **clans,** or large groups of people who are united by a common ancestor in the far past. Many Africans belong to a particular **lineage,** or larger family group with close blood ties to the same grandmother or grandfather.

Section Wrap-Up

Answer these questions to check your understanding of the entire section.

1. **Determining Cause and Effect** What are the causes of the high death rate in Africa south of the Sahara?

   _____
   _____
   _____

2. **Analyzing** Why do most city dwellers have a higher standard of living than rural people?

   _____
   _____
   _____

On a separate sheet of paper, write a paragraph explaining the role of the arts in the cultures of Africa south of the Sahara.

Chapter 20, Section 2

Chapter 21, Section 1 (Pages 572–575)
West Africa

Big Idea

Geographers study how people and physical features are distributed on Earth's surface. As you read, complete the diagram below. List key facts about Nigeria.

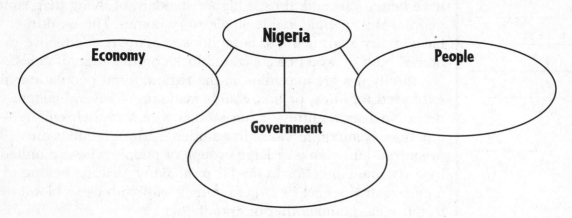

Notes | Read to Learn

Nigeria (pages 573–574)

What are Nigeria's four largest ethnic groups?

_____
_____
_____

Nigeria is a large country with the largest population in Africa south of the Sahara. Its major export is oil. The economy depends on oil sales, and Nigeria is affected whenever global oil prices change. Oil profits are used to build roads, schools, and factories.

Most of Nigeria's people have **subsistence farms,** or small plots of land where they grow food to feed their own families. Others work on larger farms that produce rubber, peanuts, palm oil, and cacao for export. The **cacao** is a tree whose seeds are used to make chocolate and cocoa. Focusing on cash crops results in not enough food being grown to feed the people, so food has to be imported.

Nigeria has more than 250 ethnic groups. The four largest are the Hausa, Fulani, Yoruba, and Ibo. The people speak many different African languages, but English is used in business and government affairs. About 50 percent of the people are Muslim, and 40 percent are Christian. The remaining 10 percent practice traditional African religions. Nigeria's cities are growing rapidly as people leave their farms in search of better jobs.

Nigeria (continued)

Nigeria gained independence from Britain in 1960. The government of Nigeria is a federal republic, but it has faced challenges in trying to build a stable democracy. Ethnic and religious differences have led to conflicts that threaten national unity.

The Sahel and Coastal West Africa (pages 574–575)

Making Inferences

Why do you think countries in the Sahel have small populations?

_____
_____
_____
_____
_____
_____
_____

The countries in West Africa can be divided into two groups based on their location. The first group lies inland in the grasslands called the Sahel. The second group is located along the Atlantic Ocean or on islands off the coast.

Five countries make up the Sahel subregion—Mauritania, Mali, Burkina Faso, Niger, and Chad. All except Mauritania are **landlocked,** meaning they do not border a sea or an ocean. Valuable deposits of uranium, gold, and oil are found in the Sahel. The lack of good transportation systems and ports limits the ability of the Sahel countries to develop these resources, however.

The Sahel receives little rainfall, so only grasses and small trees grow there. Herding livestock is a major activity. In many places, though, animals have **overgrazed** the land, or stripped it so bare that winds blow away the soil. Overgrazing and drought have led to desertification in the Sahel.

The populations in the Sahel region are relatively small compared to the rest of Africa. Many people here used to be nomads. Today, most of the Sahel people live in small towns or rural villages. The vast majority are Muslim and speak Arabic. Yet many African languages and French also are spoken here.

Coastal West Africa

Naming

Write down six countries located in coastal West Africa.

1. _____
2. _____
3. _____
4. _____
5. _____
6. _____

Coastal West Africa includes the mainland countries from Senegal to Benin along the Atlantic coast. This region also includes the Cape Verde Islands. Rain forests along the coast have been cleared for plantations, which grow palm trees, coffee, cacao, and rubber. This has led to deforestation in many areas. In search of work, rural people have migrated to port cities such as Dakar in Senegal and Accra in Ghana. As a result, the coasts are densely populated.

People in coastal West Africa belong to many ethnic groups and speak a variety of languages. They practice traditional African religions, Christianity, and Islam.

Chapter 21, Section 1

Notes | Read to Learn

The Sahel and Coastal West Africa (continued)

The countries of Liberia, Sierra Leone, and Côte d'Ivoire have experienced political unrest. In recent years, civil wars have taken many lives and destroyed their economies. The countries of Ghana, Senegal, and Benin have stable democracies, and their economies are generally prosperous.

Section Wrap-Up
Answer these questions to check your understanding of the entire section.

1. **Comparing and Contrasting** In the Venn diagram below, compare the crops grown in Nigeria with those grown in the coastal countries of West Africa.

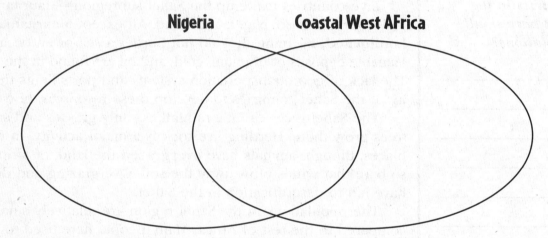

Nigeria **Coastal West AFrica**

2. **Determining Cause and Effect** Why are the people of the Sahel unable to make good use of their natural resources?

_____
_____

Descriptive Writing *In the space provided, write a paragraph describing the causes of overgrazing and the impact on the Sahel region.*

_____
_____
_____
_____

Chapter 21, Section 2 (Pages 576–582)
Central and East Africa

Big Idea

Cooperation and conflict among people have an effect on the Earth's surface. As you read, complete the chart below. List the exports of three countries in this region.

Country	Exports

Notes | Read to Learn

Central Africa (page 577)

Paraphrasing

As you read, fill in the blanks to complete this sentence.

_____
_____
and _____
_____

have prevented the Democratic Republic of the Congo from taking full advantage of its

_____
_____ .

Central Africa's seven countries are the Democratic Republic of the Congo, Cameroon, the Central African Republic, Congo, Gabon, Equatorial Guinea, and São Tomé and Príncipe.

The Democratic Republic of the Congo is a large country with mountains and savannas. One of the world's largest rain forests covers the central region and is rapidly being cleared for timber and farmland. The country has large deposits of copper, tin, and industrial diamonds, but it is unable to take full advantage of these resources. The minerals are in the center of the country, which is difficult to access because of thick forests and a lack of roads. A civil war also has hurt mining efforts.

More than 200 different ethnic groups live in the Democratic Republic of the Congo. Most Congolese dwell in rural areas, but the capital, Kinshasa, has more than 6 million people. Although African languages are spoken, French is the official language.

Gabon exports oil, manganese, uranium, and timber. Cameroon produces cacao and coffee. Equatorial Guinea, like Gabon, exports oil. São Tomé and Príncipe exports cacao, coconuts, and oil. Weak governments in Congo and the Central African Republic have kept those two countries in poverty.

Chapter 21, Section 2   145

Read to Learn

Southern East Africa (pages 578–580)

Identifying

In which countries is tourism an important industry?

_____

_____

Naming

What ethnic groups fought a long civil war in Burundi and Rwanda?

_____

_____

Tanzania's many ethnic groups have their own languages, but most people speak Swahili. A stable government and ethnic cooperation have prevented conflict in the country. Tanzania's exports include coffee and **sisal,** a plant fiber used to make rope and twine. Zanzibar, an island off Tanzania's coast, produces a spice called clove. National parks in Tanzania protect the **habitats,** or natural environments, of the country's wildlife. Serengeti National Park is a popular destination for ecotourists who come to see lions, zebras, and wildebeests.

Kenya has a developing free market economy. The capital city of Nairobi is a major business center for all of East Africa. Most Kenyans are farmers, raising corn, bananas, sweet potatoes, and **cassava,** a plant whose roots are ground to make porridge. Larger farms raise coffee and tea for export. Kenya's economy, like Tanzania's, also depends on ecotourism. The Kikuyu make up 25 percent of the population, but many other ethnic groups live in Kenya as well. Since its independence from Britain in 1963, Kenya has enjoyed prosperity and stability.

Uganda, Rwanda, and Burundi are landlocked countries in the highlands of East Africa. Subsistence farms there produce bananas, cassava, potatoes, corn, and grains. Plantations grow coffee, cotton, and tea. Since independence, all three countries have faced unrest. In the 1970s, Uganda was ruled by Idi Amin, a cruel dictator. The country today is more democratic and prosperous. Conflict between the Hutu and Tutsi ethnic groups in Rwanda and Burundi led to civil war and genocide in the 1990s. **Genocide** is the deliberate murder of a group of people because of their race or culture.

The Horn of Africa (pages 580–582)

Describing

Describe Sudan's landscape.

North

1.
2.
3.

South

The Horn of Africa is shaped like a horn that juts into the Indian Ocean. It is located in the northern part of East Africa. The region includes Sudan, Ethiopia, Eritrea, Djibouti, and Somalia.

Sudan is the largest country in Africa. It is covered by the Sahara and Nubian Desert in the north, grassy plains in the center, and fertile soil and swamplands in the south. Along the Nile River, farmers grow sugarcane, grains, dates, and cotton—the leading export. Large reserves of oil are found in the south. A 20-year civil war between Arab Muslims in the north and African Christians in the south has devastated the country.

 ## Read to Learn

The Horn of Africa (continued)

Identifying

What is the oldest independent country in Africa?

_____

What is the newest country in the Horn of Africa?

_____

Landlocked Ethiopia is covered with hot lowlands, temperate highlands, and rugged mountains. Farmers raise grains, sugarcane, potatoes, and coffee—the main export crop. About 85 percent of Ethiopians live in rural areas. The capital, Addis Ababa, is one of the largest cities in East Africa. About 80 languages are spoken, but Amharic is Ethiopia's official language. Ethiopia is the oldest independent nation in Africa. Its attempts to build a democracy have been hurt by warfare with neighboring Eritrea.

Eritrea is a small Muslim country located on the Red Sea. The newest country in the Horn of Africa, it broke away from Ethiopia in 1993. Most of Eritrea's people are farmers.

Somalia, a country shaped like the number seven, is located on the tip of the Horn of Africa. Even so, it has few natural harbors. The climate and landscape are hot and dry. Most of the people are nomadic herders, but farmers in the south grow fruits and sugarcane. Nearly all Somalians are Muslims, but they belong to different clans. Warfare among these clans began in the 1980s and continues today.

Tiny Djibouti is located at a narrow water passage linking the Red Sea and the Gulf of Aden. It is the most stable country in the Horn of Africa. Djibouti has constructed a modern port, and the country's economy is built on shipping and commerce.

Section Wrap-Up
Answer these questions to check your understanding of the entire section.

1. **Contrasting** How do Tanzania and Kenya differ from Burundi and Rwanda in regard to ethnic relations?

_____

_____

2. **Differentiating** What makes Djibouti's economy unique in the Horn of Africa?

_____

_____

 *On a separate sheet of paper, write a paragraph explaining how some countries in Central Africa have improved their economies, while others have remained poor.*

Chapter 21, Section 2

Chapter 21, Section 3 (Pages 588–592)
Southern Africa

Big Idea

Patterns of economic activities result in global interdependence. As you read, fill in the diagram below with key facts about South Africa.

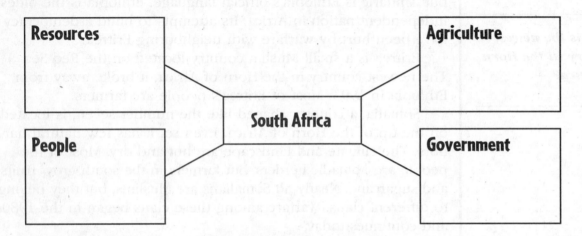

Read to Learn

Republic of South Africa (page 589)

Calculating

If you met 10 people from South Africa, how many of them would you expect to be of European descent?

_____

What percent of South Africa's population is either South Asian or has a mixed background?

The Republic of South Africa, usually called South Africa, has the most highly developed economy in Africa. It exports many minerals, including gold, diamonds, and platinum. The country also has industries, farming, and ranching. In spite of its prosperity, however, many people are poor. Rural villagers rely on subsistence farms. In the cities, industries have not grown fast enough to provide jobs for all of the people who need them.

The population of South Africa is diverse. About 75 percent of the population is made up of black ethnic groups, such as the Zulu and Xhosa. About 10 percent of the population is of European descent. These people are mainly British and Afrikaners—people who are descendants of Dutch, German, and French settlers. The remaining people are either of South Asian descent or have mixed European, Asian, and ethnic African backgrounds.

British and Afrikaner settlers formed a white-ruled South Africa in the early 1900s. They set up apartheid to control the rest of the population. Black South Africans protested apartheid, and it ended in the early 1990s. South Africa wrote a new **constitution** that gave people of all races equality. It also gave **suffrage,** or the right to vote, to all citizens aged 18 or older.

Other Southern African Countries (pages 591–592)

Inland Southern Africa

The inland, landlocked countries of southern Africa are Lesotho, Swaziland, Botswana, Zimbabwe, Zambia, and Malawi. Each has a mild climate and high plateaus. Most of their people are subsistence farmers who live in rural villages. Thousands of people are **migrant workers** who move to other areas for mining or factory work and return home only a few times each year.

Lesotho and Swaziland are located within the boundaries of South Africa. These two countries are **enclaves,** or independent territories inside another country. Both are poor countries and depend on South Africa for goods and markets.

Botswana lies directly north of South Africa. Its landscape includes swamplands and part of the vast Kalahari Desert. Its economy is based on mining and exporting diamonds and other minerals. The dry climate is not good for farming, so food is imported. Still, Botswana has a strong democracy.

Zimbabwe, northeast of Botswana, is rich in gold, copper, iron ore, and asbestos. Plantations grow coffee, cotton, and tobacco. In recent years, the government has tried to give European-owned lands to Africans. This has led to disorder, violence, and widespread shortages.

Zambia relies on copper exports for most of its income. People in cities work in mining and service industries. Villagers grow corn, rice, and other crops on subsistence farms.

To the east of Zambia is Malawi, which has wetlands, lakes, mountains, and forests. Malawi exports tobacco, tea, and sugar. Tourism also is a vital part of its economy, with visitors traveling to see wildlife in the country's national parks.

Coastal and Island Countries

Angola and Namibia lie along the Atlantic Ocean. Hilly grasslands and rocky deserts make up their landscapes. Angola has major oil reserves. Namibia mines diamonds, copper, gold, and zinc. In spite of all these natural resources, most people in Angola and Namibia are poor. They either herd livestock or live as subsistence farmers.

Across the continent, Mozambique borders the Indian Ocean. Civil war and famine have slowed its development. Recently, however, foreign investors have shown an interest in the country.

Madagascar, Comoros, Mauritius, and Seychelles are island countries in the Indian Ocean. Their populations include Asians as well as Africans. They all have economies that depend on agriculture. Mauritius also has a growing banking industry, and Seychelles has a robust tourist industry.

Explaining

Write a sentence explaining what an enclave is and stating which countries are enclaves.

_____
_____
_____
_____
_____

Naming

What are three coastal countries in southern Africa?

_____
_____
_____

Section Wrap-Up

Answer these questions to check your understanding of the entire section.

1. **Theorizing** Why is having suffrage significant to black South Africans after the end of apartheid?

   _____
   _____

2. **Defining** What does it mean to say that a country is landlocked? Name the landlocked countries in Southern Africa.

   _____
   _____
   _____
   _____

In the space provided, write a paragraph explaining the job and life of a migrant worker.

_____
_____
_____
_____
_____
_____
_____
_____
_____
_____
_____
_____
_____
_____

Chapter 22, Section 1 (Pages 610–615)
Physical Features

Big Idea

Geographic factors influence where people settle. As you read, fill in the diagram below with key facts about South Asia's physical environment.

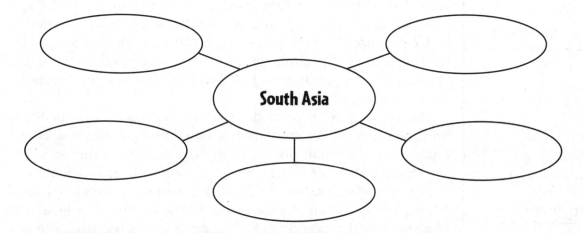

 **Read to Learn**

Landforms and Resources (pages 611–613)

Describing

Write a sentence describing how the subcontinent of South Asia was formed.

_____
_____
_____
_____
_____
_____
_____

The seven countries of South Asia include India, Pakistan, Bangladesh, Nepal, Bhutan, Sri Lanka, and Maldives. Most of these countries are on the Indian subcontinent. A **subcontinent** is a large landmass that is a part of a continent.

Three mountain systems—the Hindu Kush, the Karakoram, and the Himalaya—separate the subcontinent from the rest of Asia. Mount Everest, the world's tallest mountain at 29,028 feet high, rises in the Himalaya of Nepal.

The Himalaya protected Nepal and Bhutan from outsiders until the 1900s. However, people entered other parts of South Asia through mountain passes in the Hindu Kush. For centuries, caravans and armies marched through Khyber Pass and into India.

Scientists believe that the subcontinent of South Asia was once part of Africa's landmass. The subcontinent broke away and drifted across the Indian Ocean. When it collided with Asia, the force pushed up the mountains. Tectonic plate movements are still pushing up the mountains and causing earthquakes.

Wide, fertile plains lie south of the mountains. Three great rivers—the Indus, the Ganges, and the Brahmaputra—water the plains.

Chapter 22, Section 1

Landforms and Resources (continued)

Specifying

Write the natural resources of the region in the appropriate box.

Mineral Resources

Energy Resources

People depend on these rivers for farming, transportation, and trade. The Ganges and Brahmaputra Rivers join in Bangladesh and form the world's largest **delta,** or soil deposit at the mouth of a river.

At the base of the subcontinent are two chains of eroded coastal mountains, the Eastern Ghats and the Western Ghats. The Deccan Plateau, a highland area between these mountains, is dry because the mountains block the rain. Farther south is the lush and green Karnataka Plateau. It receives the rain instead. Spices grow on plantations here, and elephants roam through the dense rain forests.

Sri Lanka and Maldives are island countries. Teardrop-shaped Sri Lanka lies off the southeast coast of India. It has coastal lowlands and a small pocket of highlands in the interior.

Maldives lies off India's western coast and includes more than 1,300 islands. Many of these are **atolls,** or circular-shaped islands made of coral. In the center of the atoll is a shallow body of water called a **lagoon.** The lagoon is protected from the sea by the outer ring of the island. People live on only about 200 of the Maldives islands.

South Asia is not rich in natural resources. Most of the people farm small plots of land or tend livestock. India has most of the region's mineral resources. These include iron ore, manganese, and chromite. Limestone is quarried in Pakistan. Energy resources in the region include natural gas and coal. Another source of energy is hydroelectric power, which is created from fast-flowing mountain rivers.

Environmental Concerns (pages 614–615)

Listing

List three reasons supplies of freshwater are scarce in South Asia.

_____

_____

_____

South Asia is one of the most densely populated regions on Earth. Twenty percent of the world's people live on 3 percent of the world's land, and the population is increasing.

Freshwater is scarce in South Asia. The region experiences long dry seasons. Farmers often use wasteful irrigation techniques. Old, leaky pipes in cities also waste water. Some countries are tapping underground aquifers for freshwater. In urban areas, however, salt water enters the aquifers as freshwater is pumped out.

Water pollution is widespread. The Ganges River is one of the most polluted waterways in the world. Sewage, runoff from factories and fertilizers, and human waste flow into urban areas.

Chapter 22, Section 1

 **Read to Learn**

Environmental Concerns (continued)

Differentiating

As you read, complete these sentences.

Air pollution in cities is caused by

_____

_____.

Air pollution in rural areas is caused by

_____

_____

_____.

Only a small part of South Asia has forests. Most trees were cleared centuries ago. Of the forests that remain, many are being cut down for building materials and fuel. Deforestation has led to erosion and flooding, and some countries are taking steps to limit the clearing of forests.

Another concern is air pollution. The number of cars in the cities has risen rapidly in recent years. All of these vehicles release exhaust fumes, which makes the air dangerous to breathe. In rural areas, people burn wood, kerosene, charcoal, and animal dung, which releases smoke and chemicals. Many people develop breathing problems. The air pollution in South Asia and Southeast Asia is so bad that it has formed a brown cloud of chemicals, ash, and dust over the Indian Ocean. Scientists think the cloud is changing the region's climate and rain patterns.

Section Wrap-Up *Answer these questions to check your understanding of the entire section.*

1. **Identifying** Name the three great rivers of South Asia. Why are they important?

   _____

   _____

2. **Drawing Conclusions** Would people be more likely to live on the Deccan Plateau or the Karnataka Plateau? Why?

   _____

   _____

Descriptive Writing *In the space provided, describe an atoll.*

_____

_____

_____

Chapter 22, Section 1

Chapter 22, Section 2 (Pages 617–620)
Climate Regions

Big Idea

The physical environment affects people. As you read, complete the outline below to summarize the monsoon cycle.

I. First Main Heading _____

   A. Key Fact _____

   B. Key Fact _____

II. Second Main Heading _____

   A. Key Fact _____

   B. Key Fact _____

III. Third Main Heading _____

   A. Key Fact _____

   B. Key Fact _____

Notes | Read to Learn

Monsoons (pages 618–619)

Labeling

Label South Asia's seasons.

_____

October
November
December
January
February

_____

March
April
May

_____

June
July
August
September

South Asia has three distinct seasons—hot, wet, and cool. Seasonal winds called **monsoons** determine when these seasons begin and end.

Monsoons follow a yearly pattern. From October to late February is the cool season. Dry monsoon winds blow from the north and northeast. From late February to June is the hot season, and the air is heated by warm temperatures. The air then rises and causes a change in wind direction. As a result, moist ocean air moves in from the south and southeast, bringing with it the monsoon rains. This wet season lasts from June or July through September. When the monsoons sweep over the Ganges-Brahmaputra delta, the Himalaya block them from going north. So the rains turn west and fall on the Ganges Plain.

Natural Disasters

The high temperatures of the hot season and the rains of the wet season have good and bad effects on South Asia. Farmers are able to grow crops such as rice in the high temperatures. However, the heat causes water to evaporate and dries out the soil. Similarly, the rains allow crops to grow in Bangladesh and

Monsoons (continued)

Labeling

Circle the three terms that mean "an intense tropical storm with high winds and heavy rains."

the Ganges Plain. Yet areas outside the monsoon's path—the Deccan Plateau and western Pakistan, for example—receive little or no rainfall and can become scorched by drought. Too much rain can cause floods, which kill people and animals, ruin crops, destroy homes, and wipe out roads.

Another kind of weather disaster that strikes South Asia is a **cyclone,** an intense tropical storm with high winds and heavy rains. Cyclones are like hurricanes in the Atlantic and typhoons in the north Pacific. These storms can be followed by deadly tidal waves that surge from the Bay of Bengal.

Climate Zones (pages 619–620)

Finding the Main Idea

What is the main idea of this subsection?

_____
_____
_____
_____
_____

Determining Cause and Effect

Explain why the Deccan Plateau has a dry climate.

_____
_____
_____
_____
_____

Location, landforms, and monsoons affect the climate zones in South Asia. In much of the region, the climate is tropical. In the north and west, the climate can range from cold in the Himalaya to intensely hot in the deserts around the Indus River.

Tropical Areas

A tropical dry climate is found in much of south central India. Grasslands and deciduous forests are lush and green in the short wet season but turn brown in the long dry season.

A tropical wet climate is found in Bangladesh and southern Sri Lanka. Temperatures are warm throughout the year, and plenty of rain falls. Most of Bangladesh receives 100 inches of rain per year. One of the wettest places on Earth is the city of Cherrapunji in northeastern India. It averages 450 inches of rain per year.

Dry and Temperate Climates

Some areas do not benefit from the monsoons. For example, the land is dry and windswept along the lower Indus River. Farmers in this area use irrigation to water their wheat and other crops.

The sand dunes and gravel plains of the Thar Desert are located east of the Indus River. Surrounding much of this desert is a steppe. The steppe is partly dry grassland with few trees. Another steppe area crosses the Deccan Plateau. The mountains of the Western Ghats block the rain from reaching the central Deccan, so it is dry.

The Ganges Plain lies north of the Deccan Plateau. The climate there is humid and subtropical. Temperatures are high, with muggy summers but dry winters.

Chapter 22, Section 2

Climate Zones (continued)

Highlands

The Himalaya, Karakoram, and Hindu Kush mountain ranges rise along the northern border of South Asia. This section of the region has a highland climate zone. The temperature is always below freezing at elevations above 16,000 feet. Snow never melts at these elevations, and little vegetation can grow there. The climate turns more temperate farther down the mountain slopes.

Section Wrap-Up

Answer these questions to check your understanding of the entire section.

1. **Categorizing** Complete this table. Identify the positive and negative effects of the high temperatures of the hot season and the rains of the wet season.

High Temperatures		Rains	
Positive	Negative	Positive	Negative
A.	B.	C.	D.

2. **Locating** What parts of South Asia receive the most rainfall? How much rain do they get?

_____
_____

In the space provided, write a paragraph identifying at least four climate zones in South Asia. Include facts about each zone.

_____
_____
_____
_____

Chapter 23, Section 1 (Pages 626–632)
History and Governments

Big Idea

Geography is used to interpret the past, understand the present, and plan for the future. As you read, list key events and dates in this region's history.

```
┌──────────┐   ┌──────────┐   ┌──────────┐
│          │   │          │   │          │
└──────────┘   └──────────┘   └──────────┘

◄─────────────────────────────────────────►

┌──────────┐   ┌──────────┐   ┌──────────┐
│          │   │          │   │          │
└──────────┘   └──────────┘   └──────────┘
```

Notes | Read to Learn

Early History (pages 627–629)

Specifying

Write three possible causes of the decline of the Indus Valley civilization.

1. _____
2. _____
3. _____

South Asia's first cities—Harappa and Mohenjo Daro—were built in the Indus River valley by 2500 B.C. The cities were well planned, with carefully laid-out streets, ceremonial gateways, and buildings for grain storage. These cities also had plumbing and sewers. Farming, small industries, and trade brought wealth to the Indus Valley. The people made copper and bronze tools, clay pottery, and cotton cloth. In addition, the people had a writing system.

Between 1700 B.C. and 1500 B.C. the Indus civilization declined. Historians think earthquakes and floods might have damaged the cities. In addition, the Indus River might have changed its course.

Nomadic herders called Aryans settled in parts of northern South Asia about 1500 B.C. They developed a spoken language called Sanskrit. The Aryans passed on their religious teachings by word of mouth. Later, when Sanskrit became a written language, the Aryans' traditions were recorded in holy writings called the Vedas.

The Aryans were organized into four broad social groups called **varnas.** Priests were at the highest level, followed by warriors, farmers, and then servants. Over time, a caste system

Early History (continued)

Organizing

Add facts about Hinduism to the diagram.

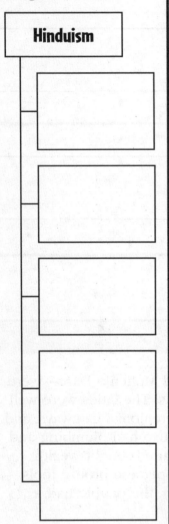

developed. A **caste** is a social group that someone is born into and cannot change. A caste determines a person's job and possible marriage partner.

The religions of Hinduism and Buddhism arose in South Asia. Hindus worship many deities, which are all part of one eternal spirit called Brahman. Hindus believe all souls want to be reunited with Brahman. To do so, a soul goes through a process called **reincarnation,** or being born into a new body after dying. To make sure their next lives are better, Hindus believe they must perform their caste's duty, or **dharma.** The effects of how a person lives are known as **karma.** Thus, if Hindus do their duty, they will have good karma.

Buddhism arose in South Asia in the 500s B.C. It was founded by a prince named Siddhartha Gautama, who was in search of the truth. He became known as the Buddha, or "Enlightened One." The Buddha taught the Eightfold Path, which people could follow to escape suffering and reach **nirvana,** or a state of endless peace and joy. Buddhism won followers among the poor and eventually spread to East Asia and Southeast Asia. In India, Buddhist ideas merged with Hinduism, which remained the major religion.

In the 300s B.C., the Mauryan Empire arose in South Asia. Aśoka, the most famous Mauryan ruler, increased trade and culture throughout the subcontinent. About 260 B.C., he became a Buddhist and dedicated his life to peace.

About A.D. 320, Chandragupta I established the Gupta Empire in northern India. This Hindu empire increased trade with other parts of the world. Science, mathematics, medicine, and the arts thrived. Gupta mathematicians developed the numerals 1 to 9 that we use today.

Muslim warriors known as Moguls established an empire in South Asia in the early 1500s, which lasted until the early 1700s. Akbar, the greatest Mogul ruler, added new lands, lowered taxes, and treated his people fairly.

Modern South Asia (pages 630–632)

During the 1600s, English traders from the British East India Company built trading posts and forts along the coast of India. By the mid-1800s, the company had colonized much of India. European ideas and practices were introduced, which the local people resented. In 1857 Indian soldiers rebelled. The rebellion was put down, and the British government took direct control

Modern South Asia (continued)

Describing

What was the British East India Company?

_____

_____

Analyzing

Why was India divided into two countries?

_____

_____

_____

_____

of India. The British brought some positive changes to the country, but taxes were heavy, local industries suffered, and famines occurred.

By the early 1900s, independence movements were spreading across South Asia. In India, Mohandas Gandhi protested British rule through nonviolent **civil disobedience**—the refusal to obey unjust laws using peaceful protests. Gandhi and his followers **boycotted,** or refused to buy, British goods.

After World War II, Britain realized the need to give South Asia its independence. Bitter divisions existed between Hindus and Muslims, however. In 1947 India was divided into two countries. The Hindu areas became India, and the Muslim areas became East and West Pakistan. In 1971 East Pakistan declared its independence and became the country of Bangladesh. Britain gave Ceylon (now Sri Lanka) its independence in 1948. Maldives became independent in 1965. Nepal and Bhutan have always been free of European rule.

Religious and political conflicts continue to trouble South Asia. Pakistan and India both claim and have fought wars over Kashmir, a region in the Himalaya and Karakoram. Both countries have nuclear weapons, which worries other countries. A civil war that began in 1983 between the Sri Lankan government and ethnic Tamils continues. Democratic groups forced Nepal's king to give up many political powers in 2006, but communist rebel forces still control large areas of Nepal's countryside.

Section Wrap-Up

Answer these questions to check your understanding of the entire section.

1. Defining What is a caste system, and how did one emerge in South Asia?

_____

_____

2. Explaining How did South Asia come to be dominated by the British?

_____

_____

Choose a law that you feel is unjust. On a separate sheet of paper, write a letter to the editor to convince people to protest that law through civil disobedience.

Chapter 23, Section 1

Chapter 23, Section 2 (Pages 638–644)
Cultures and Lifestyles

Big Idea

The characteristics and movements of people impact physical and human systems. As you read, fill in the chart below with examples of culture in South Asia.

Element of Culture	Example
Religion	
Arts	
Daily Life	

Notes / Read to Learn

The People of South Asia (pages 639–640)

Stating

What are five challenges facing South Asia's slums and two challenges facing rural areas?

Cities:

1. _____
2. _____
3. _____
4. _____
5. _____

Rural Areas:

1. _____
2. _____

Nearly 1.5 billion people live in South Asia. The population grew dramatically during the last century. Improved health care lowered death rates, and birthrates remained high. The region is densely populated as well. Bangladesh is the most densely populated country, with 2,584 people per square mile.

The population is not evenly distributed throughout the region. River valleys are crowded, while desert areas have few people. Two-thirds of the people live in rural areas, but the region also has large and growing cities. Mumbai (Bombay), India, has more than 19 million inhabitants.

City streets in South Asia are crowded with people, animals, carts, bicycles, and cars. Skyscrapers and apartments reveal wealth and a growing middle class, but poverty is widespread. Many urban dwellers have inadequate housing or are homeless. People living in slums face unemployment, pollution, disease, crime, and a lack of clean water.

In Nepal and Bhutan, the mountainous terrain limits farming. In other rural areas, overcrowding has reduced the size of farmers' plots, so they cannot grow enough food to feed their families. Some villages lack electricity and safe drinking water.

The People of South Asia (continued)

Identifying

How many official languages does India have?

_____

South Asia has many different ethnic groups. The Sherpa of Nepal are a farming people known for their climbing ability. They often are hired to lead mountain climbing expeditions. The people of South Asia speak 19 major languages and hundreds of local dialects. India alone recognizes 15 official languages, although Hindi is the primary one. Urdu is spoken in Pakistan, and people in Bangladesh speak Bengali. English is widely spoken in areas that were once under British rule.

Religion, the Arts, and Daily Life (pages 641–644)

What is the ratio between Hindus and Muslims in India?

_____

_____

Making Connections

How has religion influenced the arts in South Asia?

_____

_____

_____

_____

_____

Hinduism is the most widely practiced religion in South Asia. In India, Hinduism helps unify the large population. It impacts the daily life of about 800 million Indians, all of whom share the same sacred writings and perform common rituals.

The second-largest religion in the region is Islam. Pakistan, Bangladesh, and Maldives were founded as Islamic countries. Islam has 140 million followers in India. Muslims and Hindus sometimes clash, but India's secular government protects all citizens.

Buddhism is a major religion in Sri Lanka, Nepal, and Bhutan. *Dzongs,* or Buddhist centers of prayer and study, have shaped the arts and culture of Bhutan. Other religions practiced in the region include Sikhism, Jainism, and small pockets of Christianity.

The Arts

The arts in South Asia are strongly influenced by religion. Sacred writings have inspired painters. Hindu, Buddhist, and Sikh temples hold elaborate carvings and sculptures of Hindu deities or the Buddha. Muslims also built beautiful mosques, forts, and palaces. The famous Taj Mahal in Agra, India, was built by a Muslim ruler as a tomb for his wife.

Much of South Asian literature is rooted in religion. The *Mahabharata,* one of India's sacred texts, was written about 100 B.C. In a section known as the Bhagavad Gita, the deity Krishna explains that it is noble to do one's duty even when it is difficult. South Asians also tell stories through plays and dance. India's traditional dances are based on Hindu themes.

Classical Indian music usually features a long-necked instrument called the **sitar.** It has 7 strings on the outside and 10 inside the neck. Contemporary music, such as rock music popular in Pakistan, shows Western influences.

Moviemaking is a booming business. Mumbai, nicknamed "Bollywood," releases hundreds of movies every year.

Religion, the Arts, and Daily Life (continued)

Explaining

How has religion influenced people's diets in South Asia?

_____
_____
_____
_____

Daily Life

Family is the center of life in South Asia. Marriage is viewed as the joining of two families. In India and Pakistan, extended families often live together in one house.

In villages, people live in homes with mud walls and thatched roofs. In Bangladesh, homes are built on stilts to protect against frequent floods. Urban middle- and upper-class people live in comfortable houses or apartments. Millions of poor people crowd into cheap apartments or flimsy shacks, or they sleep on the streets.

Some city dwellers dress in western-style clothing, but many prefer traditional garments. For example, Indian women wear a long, rectangular piece of cloth, called a **sari,** which is draped around the body.

Religion impacts the foods eaten in South Asia. Hindus do not eat beef, Muslims do not eat pork, and Jains—among others—avoid all meat. Vegetarian cooking and spicy sauces are popular.

Two of the most popular sports in the region are field hockey and cricket. Both were introduced by the British during the colonial era.

Section Wrap-Up

Answer these questions to check your understanding of the entire section.

1. **Categorizing** Which religions have the largest and second-largest number of followers in South Asia? What other religions are practiced?

   _____
   _____

2. **Describing** Describe the different types of housing in South Asia.

| Villages | Cities ||
	Upper and Middle Class	Poor
A.	B.	C.

On a separate sheet of paper, write a paragraph explaining the importance of Hinduism in India, and the secular government's role in the clashes between Muslims and the Hindu majority.

Chapter 24, Section 1 (Pages 652–657)
India

Big Idea

Patterns of economic activity result in global interdependence. As you read, list key facts about India's economy in the diagram below.

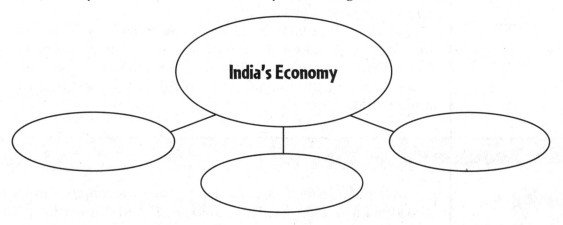

Notes — Read to Learn

India's Government (pages 653–654)

Complete this diagram with details of India's national government.

Executive Branch
▸ _____
▸ _____

Legislative Branch
▸ _____
▸ _____

Judicial Branch
▸ _____

India, with more than 1 billion people, is the world's largest democracy. Like the United States, India is a federal republic. Power is shared between the national government—located in New Delhi, the capital—and 28 state governments. The states have their own duties, such as providing police protection. If a state law conflicts with a national law, the national law is followed.

India's 28 states vary widely in size and population. Some states are made up of one ethnic group. Those states allow the group to focus on their particular needs. India also has seven union territories that the national government controls. These include some offshore islands and several cities.

India's national government has executive, legislative, and judicial branches. This separation of powers means that each branch has its own duties that the other branches cannot interfere with.

India has a president who is a ceremonial head of state. The chief executive with real power is the prime minister who heads the government and sets policy. Jawaharlal Nehru, elected in 1947, was India's first prime minister. His daughter, Indira

India's Government (continued)

Specifying

Which legislative house is more democratic?

_____

_____

Gandhi, also was prime minister until she was assassinated in 1984.

The legislative branch makes the laws. India's legislature is made up of two houses—the larger People's Assembly elected by voters, and the smaller Council of States chosen by the prime minister or state legislatures.

The judicial branch is led by the Supreme Court. It interprets laws to be sure they agree with the country's constitution. The Indian constitution, which identifies citizens' rights and duties, is one of the longest and most detailed constitutions in the world.

India's Economy (pages 656–657)

Listing

What are India's primary crops?

1. _____
2. _____
3. _____
4. _____
5. _____
6. _____

Differentiating

Underline examples of cottage industries. Circle examples of factory-based industries.

After India gained independence, the government ran much of the country's industry. The economy slowed during the 1970s. In response, India began moving to a free market economy as a way to boost growth. Businesses became privately owned, and foreign investment increased. Today India has one of the most rapidly growing economies in the world. Still, not enough jobs exist for the huge population, and many people remain poor.

Agriculture and Related Industries

More than half of India's land is used for agriculture, and 75 percent of the people are farmers. In the 1970s, India benefited from the **green revolution**—changes that modernized agriculture and increased food production. Farmers raise rice, wheat, cotton, tea, sugarcane, and jute. **Jute** is a plant used to make rope, burlap bags, and carpet backing.

India also has rich deposits of coal, iron ore, manganese, bauxite, and diamonds. Fishing is another important industry. India's investments in oceangoing ships and processing plants have increased fish exports.

Manufacturing and Services

India has **cottage industries**—people work in their own homes and use their own equipment to make products. Cottage industries produce pottery, cloth, and metal and wooden goods. India also has factory-based industries. Textile factories employ the most workers and produce natural and synthetic fabrics. Other industries include food processing plants and factories that make steel, locomotives, trucks, and chemicals.

The service sector is the fastest-growing part of India's economy. Computer software services are especially booming in

India's Economy (continued)

Which two Indian cities are software service centers?

_____

_____

southern Indian cities, such as Hyderabad and Bangaluru. Many software developers and technical support specialists actually work for American companies. American businesses use **outsourcing,** which means that they hire overseas workers to do certain jobs. India attracts outsourcing business because wages there are low, and many workers are educated, skilled, and fluent in English.

India also has a large number of doctors, scientists, and engineers. Some of these professionals perform outsourced work as well. For example, they research, write, and do other jobs for American companies.

Section Wrap-Up

Answer these questions to check your understanding of the entire section.

1. **Defining** How is India's federal republic organized?

   _____

   _____

2. **Evaluating** Why are many Indians poor despite the growing economy?

   _____

   _____

In the space provided, write a paragraph explaining the role of outsourcing in the Indian economy.

_____

_____

_____

_____

_____

_____

Chapter 24, Section 1

Chapter 24, Section 2 (Pages 660–664)
Muslim Nations

Big Idea

All living things are dependent on one another and their surroundings for survival. As you read, compare and contrast the economies of Pakistan and Bangladesh in the Venn diagram below.

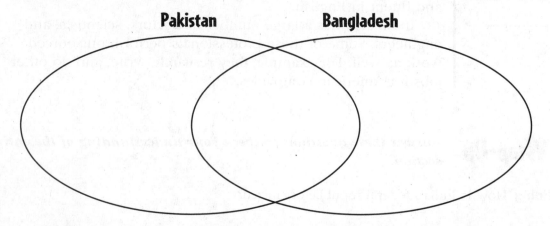

Notes | Read to Learn

Pakistan (pages 661–662)

Specifying

Which parts of Pakistan's economy are still under government control?

_____

_____

_____

A long, wide country, Pakistan is situated between Afghanistan, Iran, and India. Northern Pakistan is covered with mountains. In the south, the Indus River valley provides fertile land.

The People

Pakistan has more than 160 million people, making it one of the world's most populous countries. The death rate has declined, but the birthrate is high, so the population continues to grow. Islam is a common bond for Pakistanis. Still, many different ethnic groups live here, each with its own language, territory, and identity.

The Economy

In the 1970s, Pakistan's businesses were **nationalized,** or placed under the control of the government. Since the 1990s, many businesses have become privately owned. The government still manages banks, hospitals, and transportation, however. Manufacturing and service industries are important, but about half the people farm. A large irrigation system waters sugarcane, wheat, rice, and cotton. Cotton cloth and clothing are among

Pakistan (continued)

Naming

Over which region have Pakistan and India fought two wars?

_____

Pakistan's major exports. People also work in cottage industries, producing metalware, carpets, and pottery.

Although the economy has grown, too few jobs exist for all the people. As a result, most Pakistanis are poor. Millions get temporary work in other countries and send money back to Pakistan to support their families.

Government and Foreign Relations

Pakistan is a federal republic, but democracy is limited. The military has often taken power and forced elected leaders out of office. In 2001 Pakistan's president helped the United States in its struggle against terrorism.

Pakistan has conflicts with its neighbor, India. The countries have fought two wars for control of the territory of Kashmir. Tensions rose in 1998 when both countries successfully tested nuclear weapons. Since then, Pakistan and India have agreed to a cease-fire and have moved toward greater cooperation.

Bangladesh (pages 663–664)

Summarizing

Give three reasons Bangladesh has so much poverty.

_____
_____
_____

Explaining

Why can farmers in Bangladesh plant crops three times a year?

_____
_____
_____

Bangladesh is one of the most densely populated countries in the world. Its large population and limited resources have caused it to struggle since it became independent in 1971.

The People

Most of Bangladesh's 144 million people are Muslim. They also are poor. About 75 percent live in rural areas, but many are moving to crowded cities to find jobs. Dhaka is the capital and major port.

Bangladesh faces threats from natural disasters. Monsoon rains cause the Brahmaputra and Ganges Rivers to overflow. Cyclones also cause flooding. The floods kill people, destroy crops, and cause food shortages. The lack of food leads to malnutrition for many people.

The Economy

Most people in Bangladesh work as farmers. Plentiful water, fertile soil, and a warm climate make it possible for farmers to plant and harvest crops three times a year. The major crop is rice. Other crops include sugarcane, jute, wheat, and tea. Still, the farmers cannot produce enough crops to feed all the people of Bangladesh. They lack modern tools, and they use outdated farming practices.

Chapter 24, Section 2

Notes | Read to Learn

Bangladesh (continued)

Inferring

Why would Bangladesh want to maintain friendly relations with India?

_____
_____
_____

A major export of Bangladesh is clothing. Nearly 2 million people, mostly women, work in clothing factories. Another profitable industry is **ship breaking**—bringing ashore and tearing apart large ships that are no longer usable. The scrap metal from the old ships is sold for steelmaking or construction projects. Bangladesh also has large reserves of natural gas, but the government has not decided whether to use those reserves to meet the country's own energy needs or to export natural gas to earn money.

Surrounded by India on three sides, Bangladesh tries to cooperate with its larger neighbor. Tensions sometimes arise over the Ganges River, which both countries share. Recently, Bangladesh has become a main supplier of soldiers for United Nations peacekeeping missions.

Section Wrap-Up

Answer these questions to check your understanding of the entire section.

1. **Identifying** What type of government does Pakistan have?

   _____
   _____

2. **Explaining** Why do many people suffer from malnutrition in Bangladesh?

   _____
   _____

In the space provided, write a paragraph explaining Pakistan's relationship with India.

_____
_____
_____
_____

Chapter 24, Section 3 (Pages 665–668)
Mountain Kingdoms, Island Republics

Big Idea

Cooperation and conflict among people have an effect on the Earth's surface. As you read, list the issues that three countries in this section face. Then explain the issues' significance.

Issue	Significance
1.	
2.	
3.	

Notes | Read to Learn

Nepal and Bhutan (pages 666–667)

Determining Cause and Effect

What has resulted from farmers clearing the land in Nepal?

_____
_____
_____
_____

The Himalaya, located in northern Nepal, include 8 of the world's 10 highest mountains. Nepal's landscape also includes hills, valleys, and a fertile river plain. The capital and only major city is Kathmandu. The majority of Nepal's people live in rural villages. Buddhism is practiced, but Hinduism is the official religion.

Nepal's farmers cultivate small patches of rice and other crops. They have cleared forests for more farmland, which causes erosion and terrible flooding in the valleys. Tourism and trade help the economy. Once isolated by the mountains, today the country has roads and plane service to India and Pakistan. Nepal exports clothing and carpets. It imports gasoline, machinery, and **consumer goods,** or products that people buy for personal use. Recently, Nepal's king has been in a power struggle with pro-democracy groups and communist rebels. The political instability has kept Nepal very poor.

Bhutan

Tiny Bhutan lies east of Nepal. The peaks and forested foothills of the Himalaya cover much of the country. Plains and river valleys are found along Bhutan's southern border with India.

Read to Learn

Nepal and Bhutan (continued)

Speculating

Why do you think India helped Bhutan build hydroelectric plants?

_____
_____
_____

The mountains hinder travel into Bhutan. With little trade, the economy is struggling. Most people are subsistence farmers who live in remote villages. Now that some roads connect Bhutan with the outside world, tourism is a growing industry. The government limits the number of tourists to protect Bhutan's culture, however. India has helped Bhutan build hydroelectric plants to create electricity.

Most people are Buddhists who belong to the Bhutia ethnic group. The smaller, mostly Hindu, Nepali ethnic group complains of discrimination during the years of rule by Buddhist kings. In recent years, Bhutan has moved toward democracy.

Island Republics (pages 667–668)

Calculating

Create a circle graph that summarizes Sri Lanka's ethnic population.

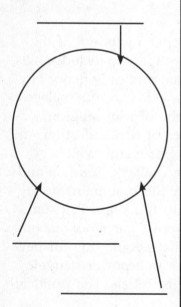

Sri Lanka is an island republic off the southeastern coast of India in the Indian Ocean. Much of the country is covered with rolling lowlands and white sandy beaches. Highlands rich in wildlife rise in the center of the island.

Sri Lankans have farmed for many years. Crops such as rice grow in the lowland areas. Plantations of rubber trees, coconut palms, and Ceylon tea are found in the highlands.

The economy is becoming more industrialized. Factories produce textiles, fertilizers, cement, leather goods, and wood products. Sri Lanka also exports sapphires, rubies, and other gemstones.

Two major ethnic groups live in Sri Lanka. The Sinhalese make up about 74 percent of the population. They are mostly Buddhist and live in the south and west. About 17 percent of the people are Tamils. They live in other parts of the country and are mainly Hindu.

The Tamils and Sinhalese have been fighting a civil war since 1983. The Tamils believe that the Sinhalese have treated them unfairly. Their goal is to establish an independent Tamil nation in northern Sri Lanka. Thousands have died in this war.

In December 2004, an earthquake on the eastern edge of the Indian Ocean launched a **tsunami,** or huge ocean wave. The tsunami struck Sri Lanka, killing more than 30,000 people and leaving 850,000 homeless.

Maldives

Maldives is made up of about 1,200 coral islands located southwest of India in the Indian Ocean. The highest elevation on the islands is 6 feet above sea level. About 360,000 people

 # Read to Learn

Island Republics (continued)

Listing

List three industries in the Maldives.

_____

_____

_____

live in the Maldives, with approximately 80,000 living in the capital, Male. Most of the people are Muslims.

With sandy soil covering the islands, farmers can grow only a few crops. Most food is imported. Tourism is now the largest industry in Maldives. People are attracted to the palm-lined, sandy beaches and coral formations on the islands. Other important economic activities include fishing and boatbuilding.

Section Wrap-Up *Answer these questions to check your understanding of the entire section.*

1. **Identify** What two main religions are practiced in Nepal and Bhutan?

   _____

   _____

2. **Labeling** Label the Indian Ocean, Sri Lanka, India, and Maldives on the map below.

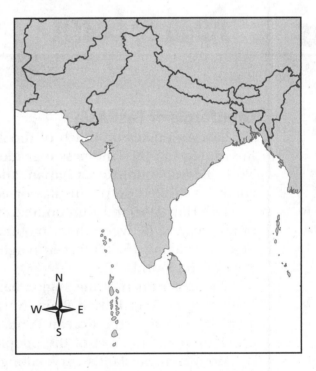

 *On a separate sheet of paper, write a paragraph summarizing the key features of Nepal's economy.*

Chapter 24, Section 3

Chapter 25, Section 1 (Pages 688–693)
Physical Features

Big Idea

Physical processes shape Earth's surface. As you read, provide an example, a location, and a key fact about each type of landform in the chart below.

Landform	Example	Location	Key Fact
Mountain			
Plateau			
Plain			
Archipelago			

Notes / Read to Learn

The Land (pages 689–690)

Illustrating

Make a sketch to show the location of the Plateau of Tibet in relation to the Himalaya and Kunlun Shan.

Landforms of East Asia

East Asia takes up much of the Asian continent south of Russia. Most of the landmass is made up of China and Mongolia. North Korea, South Korea, Japan, and Taiwan are East Asian countries located on peninsulas or islands.

The Himalaya and Kunlun Shan tower over the western part of this region. Between these mountains stands the Plateau of Tibet. Averaging 15,000 feet in height, the plateau is called the Roof of the World.

The eastern part of this region has lowlands—the North China Plain and the Manchurian Plain. Narrow lowlands also sweep along the coasts of the Korean Peninsula and Japan. These fertile areas are home to most of the people of East Asia.

Mountainous islands curve along the coast of China and the Korean Peninsula. Japan forms an **archipelago,** or chain of islands, in the Pacific Ocean. Taiwan is located off China's southeastern coast. Formed by volcanoes, these islands are part of the Pacific Ring of Fire. They experience many earthquakes and volcanic eruptions because they sit where tectonic plates meet and move.

Notes | Read to Learn

The Land (continued)

Summarizing

What are the major landforms of Southeast Asia?

_____

_____

_____

Landforms of Southeast Asia

South of China, Southeast Asia is made up of mainland peninsulas and thousands of islands. The mainland countries are Myanmar (Burma), Thailand, Vietnam, Laos, and Cambodia. Countries that are partly or entirely islands include Indonesia, Malaysia, Singapore, East Timor, Brunei, and the Philippines.

Mainland Southeast Asia is crossed by **cordilleras,** or mountain ranges that run side-by-side. Between the ranges lie fertile river plains and deltas, where most of the people live. Like East Asia, the islands of Southeast Asia are part of the Ring of Fire and experience earthquakes and active volcanoes. In 2004 an earthquake in the Indian Ocean caused a tsunami that swept over Southeast Asia's coastal lowlands. More than 300,00 people died.

Seas and Rivers (pages 690–691)

Identifying

Identify two rivers in East Asia and four rivers in Southeast Asia.

1. _____
2. _____
3. _____
4. _____
5. _____
6. _____

The countries here have long coastlines along the Indian Ocean, the Pacific Ocean, and many seas. These waterways influenced the region's history. For example, as an isolated island nation, Japan developed a unique culture.

Nearness to water affects the region's economies as well. Oceans and seas serve as trade routes. Much of the world's shipping traffic travels on the South China Sea and the Strait of Malacca. The people here also depend on the sea for food. Japan, South Korea, Taiwan, and China have the world's biggest deep-sea fishing industries.

East Asia's major rivers begin in Tibet. The Huang He (Yellow River) flows across northern China. The river carries tons of fine, yellow-brown soil called **loess.** This rich soil and river water make the North China Plain ideal for growing wheat. China's other great river is the Chang Jiang (Yangtze River). It flows through spectacular **gorges,** or canyons, and broad plains before reaching the port city of Shanghai. Half of China's rice and grain farmers depend on this river. The Chang Jiang is also a vital trade route for ships, which can travel far upriver.

Southeast Asia's major rivers include the Irrawaddy and Salween in Myanmar, and the Chao Phraya in Thailand. The Mekong River flows 2,600 miles through five countries on its way to the South China Sea. Warm temperatures, heavy rains, and fertile soil make the Mekong region ideal for growing rice. The river valleys are densely populated.

Chapter 25, Section 1

Notes | Read to Learn

A Wealth of Natural Resources (pages 692–693)

Listing

List this region's energy, mineral, and forest resources.

_____

_____

_____

_____

_____

East Asia and Southeast Asia are rich in natural resources. These resources have helped to develop the region's economies.

China has large oil deposits in the South China Sea, as well as rich coal deposits. Other oil-rich countries include Indonesia, Brunei, Malaysia, and Vietnam. Some countries generate hydroelectric power from their swift-flowing rivers. The world's largest dam—Three Gorges Dam—is being built on China's Chang Jiang to provide hydroelectric power as well as to prevent flooding.

Among the minerals found here are tin, iron ore, chromium, manganese, nickel, and tungsten. **Tungsten** is used to make lightbulbs and rockets. Gems and pearls also are plentiful in the region.

Forests provide valuable woods. Teak is harvested in Myanmar, Indonesia, and Thailand. **Teak** is a type of wood that is used to make buildings and ships because it is strong and durable. Mahogany from the Philippines is used for wall paneling and high-quality furniture.

Section Wrap-Up *Answer these questions to check your understanding of the entire section.*

1. **Identifying** In what geographic areas do most of the people of East Asia and Southeast Asia live? Why?

   _____

   _____

2. **Analyzing** How have the economies of East Asia and Southeast Asia been affected by waterways?

   _____

   _____

On a separate sheet of paper, write a paragraph describing the relative location of the countries of East Asia and Southeast Asia.

Chapter 25, Section 2 (Pages 694–698)
Climate Regions

Big Idea

Geographic factors influence where people settle. As you read, complete the diagram below. Identify the air masses that affect the climates in the region.

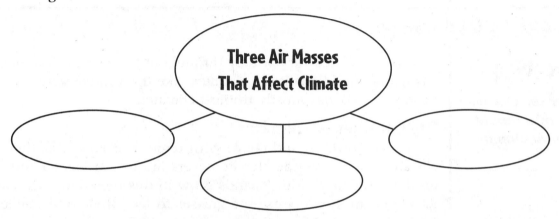

Notes | Read to Learn

Effects on Climate (page 695)

Identify three physical features that shape the climates in East Asia and Southeast Asia.

1. _____
2. _____
3. _____

Three different air masses influence the climates of East Asia and Southeast Asia. The first one brings cold, dry air from the Arctic region. The second air mass brings cool, dry air eastward across Asia from the west. The final one brings warm, moist air from the Pacific Ocean.

The winds, along with landforms, shape the climates of the region. In winter, for example, cold Arctic winds blow across Siberia. The result is lower temperatures in Mongolia and northern China.

In southern and eastern regions, monsoons are common in the summer. Warm, moist air from the Pacific Ocean blows into East Asia, bringing rain. In the winter, however, dry winds blow outward from Asia to the ocean, carrying little rain. Closer to the Equator, temperatures are warm and rain falls more evenly throughout the year.

Ocean currents also impact the climate, particularly on the region's islands. A warm-water current warms southeastern Japan. A cold current along Japan's Pacific coast brings cold winters to northern Japan.

 **Read to Learn**

Effects on Climate (continued)

Hurricane-like storms called typhoons form in the warm waters of the Tropics. They can blow across the coast of East Asia. Their high winds, heavy rains, and large waves cause much damage when they blow ashore.

Climate Zones (pages 696–698)

Describing

What are the characteristics of a humid continental climate?

_____
_____
_____
_____

Determining Cause and Effect

What factors can keep temperatures lower near the Equator?

_____
_____
_____

Climate zones vary widely throughout the region. East Asia has mainly middle latitude climates, like the United States. Southeast Asia has mostly tropical climates.

Dry Continental Climates

In the northern and far western regions of East Asia, the climate is dry. A steppe climate covers most of Mongolia and northern China. Vast grasslands grow in this mostly dry climate. Ranchers and herders use the grasses to feed their cattle, sheep, goats, and camels. Sometimes a weather pattern of a dry summer followed by a harsh winter occurs. Mongolians fear this pattern, called a *dzud*. The dry summers limit the food available for the herds, which are then too weak to survive a harsh winter.

Some areas in the region are extremely dry because surrounding mountains block moisture. In southern Mongolia and northern China lies a vast desert region called the Gobi. The name *Gobi* means "place without water" in the Mongolian language. Even less rain falls in the Taklimakan, a desert in western China.

Wet Continental Climates

Northeastern China, North Korea, and northern Japan experience a humid continental climate, which has extreme temperatures during the year. Summers in these areas are warm, and winters are cold.

The rest of East Asia and the northern part of Southeast Asia have a humid subtropical climate. Summers here are slightly warmer than in the humid continental zone, and the winters are milder. This is a good climate for growing rice.

Tropical Climates

Most of Southeast Asia is located in the Tropics and stays warm all year. In summer, the direct rays of the sun result in high temperatures. In winter, warm air from the Equator blows over the region.

Even near the Equator, however, sea breezes keep temperatures moderate along coastal areas. High altitudes also keep the temperatures low in tropical areas. Mountains on the islands of

Chapter 25, Section 2

Notes | Read to Learn

Climate Zones (continued)

Speculating

Why is deforestation particularly harmful in tropical areas?

_____
_____
_____

Borneo and New Guinea can be quite cold and are sometimes covered with snow.

Countries with a tropical rain forest climate receive heavy rainfall. These rains result in lush forests that may hold 100 different kinds of trees and thousands of flowering plants. Yet in Malaysia, Thailand, and several other countries, rain forests are rapidly being cut. Deforestation has led to **landslides,** when soil is washed down the treeless hillsides, burying villages in mud and killing villagers.

Highland Climates

The Himalaya and Kunlan Shan, the Plateau of Tibet, and some mountainous areas in Indonesia experience a highland climate. Temperatures are cool to extremely cold in these regions. Dry continental air results in dry landscapes in the highland zones.

Section Wrap-Up *Answer these questions to check your understanding of the entire section.*

1. **Explaining** What is a *dzud,* and why do Mongolians fear it?

_____
_____

2. **Determining Cause and Effect** How has deforestation affected some Southeast Asian countries?

_____
_____

 *In the space provided, write a paragraph describing some of the climate changes you would find as you travel throughout East Asia and Southeast Asia.*

_____
_____
_____
_____

Chapter 25, Section 2

Chapter 26, Section 1 (Pages 704–712)
History and Governments

Big Idea

Patterns of economic activities result in global interdependence. As you read, complete the time line below. List key events and dates in China's history.

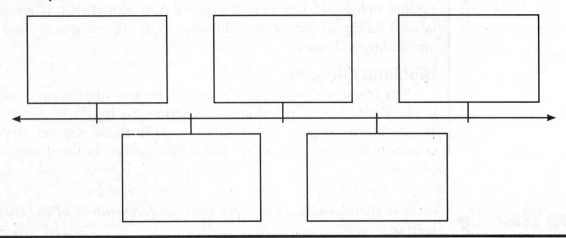

 **Read to Learn**

Asian Empires (pages 705–707)

Paraphrasing

Complete these sentences.

In China, a _____ would hold power until it was _____.

In 1211 China was invaded by _____ warriors.

Japan's _____, or military leader, was supported by _____.

Chinese civilization began in the Huang He valley more than 4,000 years ago. Until the early 1900s, emperors ruled China. A line of rulers from a single family, called a **dynasty**, would rule until another dynasty overthrew it.

During the Han dynasty (206 B.C.–A.D. 221), paper was invented and a trading network was established. Silk, tea, spices, paper, and fine clayware called **porcelain** were sent along the Silk Road as far as the Mediterranean. Later, the Tang (A.D. 618–907) and Song (A.D. 960–1279) dynasties built roads, canals, and irrigation systems within China. The population increased rapidly. The Tang and Song also invented steel, woodblock printing, gunpowder, and the compass.

In 1211 Mongol warriors invaded, eventually conquering and uniting most of China. The Ming dynasty (1368–1644) drove out the Mongols. Ming rulers hired government workers based on their skills. The people were counted in a **census** so taxes could be collected more accurately. The Ming also built the Forbidden City in Beijing, the capital, where emperors lived for more than 500 years. People staged dramas and wrote long fictional stories called **novels**.

Asian Empires (continued)

Identifying

What did the samurai code require?

_____
_____
_____

About 1200 B.C., Chinese culture spread to Korea and Japan. In the A.D. 1100s, nobles fought for control of Japan. This led to rule by a military leader called a shogun, who was supported by land-owning warriors called samurai. A code demanded that a samurai be loyal, brave, and honorable. The samurai helped shoguns rule Japan until the late 1800s.

In Southeast Asia, the Chinese ruled much of what is now Vietnam from the 100s B.C. to the A.D. 900s. During the A.D. 800s, Muslim Arab traders and missionaries settled in coastal areas of Southeast Asia. Eventually many people in these places converted to Islam. The Khmer people founded an empire over much of mainland Southeast Asia during the A.D. 1100s. They built Angkor Wat, a huge Hindu temple and royal tomb.

Modern Nations (pages 708–710)

Stating

Write a sentence explaining what a sphere of influence is.

_____
_____
_____
_____
_____

Explaining

Underline the reason Japan invaded neighboring countries in the early 1900s.

By 1500, goods from Asia became valued in Europe, and European merchants traveled to the region. The Chinese tried to limit the influence of Europeans on their culture. By the 1890s, however, European countries and Japan claimed parts of China as **spheres of influence,** or areas where a single foreign power has exclusive trading rights. The Chinese revolted against foreigners in 1911. In 1949 Chinese Communists led by Mao Zedong won control of the country and established the People's Republic of China.

Japan also tried to limit western influence. In 1854 the Japanese were forced to end their isolation. Power was taken from the shoguns and returned to the emperor. Japan adopted western technology and quickly became an industrial and military power. It needed resources, however, so it invaded neighboring countries. This expansion was one factor that led Japan to fight the United States and its allies in World War II. Japan was defeated and became a democracy in 1945.

After World War II, Korea was divided into two countries. Communist-ruled North Korea wanted to unite the countries, so it invaded South Korea in 1950, starting a war. The Korean War ended in a truce in 1953, and Korea remains divided today.

European countries dominated much of Southeast Asia during the 1800s and 1900s. They wanted the region's sugar, coffee, tea, rubber, and oil. Only Siam, known as Thailand today, remained independent. After World War II, countries in the region gained independence. Political conflicts raged throughout the region for many years.

Chapter 26, Section 1

Notes | Read to Learn

Listing

List the four Asian Tigers.

1. _____
2. _____
3. _____
4. _____

After World War II, Japan's economy was in ruins. However, Japan and the United States established close ties during the Korean War. The United States gave $3.5 billion in aid to Japan, which created an economic boom. Japan now has the second-largest economy in the world, after the United States.

The countries of South Korea, Taiwan, Singapore, and Hong Kong have been nicknamed the "Asian Tigers." They have booming economies based on exports of computers, other electronic goods, and vehicles. Singapore is a **free port,** which means that goods can be unloaded, stored, and shipped without payment of taxes. As a result, huge amounts of goods pass through the tiny country.

Communist China was slower to develop its economy because government kept control of most businesses. China began a number of free market reforms in 1979, and its economy began to grow.

Section Wrap-Up

Answer these questions to check your understanding of the entire section.

1. **Defining** What is a census, and why did the Ming take one?

   _____
   _____

2. **Summarizing** What changes occurred in East Asia and Southeast Asia after World War II?

   _____
   _____

On a separate sheet of paper, write a paragraph explaining how trade has developed in one of the countries in this region.

Chapter 26, Section 2 (Pages 714–720)
Cultures and Lifestyles

Big Idea

Culture influences people's perceptions about places and regions. As you read, complete the diagram below. List key facts about population patterns in the region.

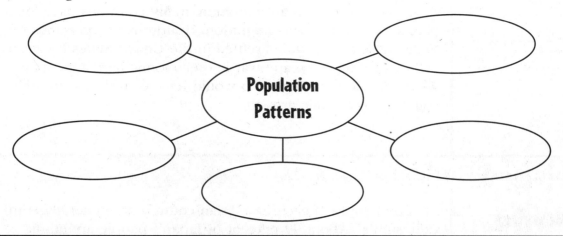

 **Read to Learn**

Population Patterns (pages 715–716)

Summarizing

Where do most people in East Asia live?

_____
_____
_____
_____

The population is increasing rapidly in East Asia and Southeast Asia, putting a burden on available resources. Agricultural countries, such as Cambodia and Laos, have high birthrates. Some countries have enacted policies to slow the population growth rate. For example, China's government encourages families to have only one child. Even so, the population continues to grow. China's greatest challenge is a shortage of jobs for its available workers. In several industrialized countries, such as Japan, South Korea, and Taiwan, the population is actually declining.

Most people in East Asia live in crowded river valleys and basins or along coastal plains. Agriculture and industry thrive in these areas because of the favorable land and climate. In contrast, the interior regions of East Asia are sparsely populated. These areas have mountains and little vegetation.

Like East Asia, the people in Southeast Asia are not evenly distributed. Large numbers of people live crowded together in small areas. The population density of Singapore, for example, is nearly 18,000 people per square mile.

Population Patterns (continued)

Identifying

Identify four cities in Japan that form a megalopolis.

1. _____
2. _____
3. _____
4. _____

Nearly 60 percent of the people live in rural areas. Recently, however, many people have migrated to cities in search of better jobs. Some cities in the region are among the largest in the world. In Japan, the cities of Tokyo, Osaka, Nagoya, and Yokohama are so close together that they form a supersized urban area called a **megalopolis.** Their combined population totals 5 million people.

In addition to urban migration, many countries have lost people due to overseas migration. Hundreds of thousands of Vietnamese and Laotians settled in the United States between 1975 and 1990 to escape war and economic hardship. Many were educated workers who would have contributed to their countries' economic growth.

People and Cultures (pages 718–720)

Specifying

What is China's official language?

_____

Naming

Name seven religions practiced in the region.

1. _____
2. _____
3. _____
4. _____
5. _____
6. _____
7. _____

The people in each East Asian country are generally ethnically similar. About 99 percent of Japan's people are ethnic Japanese who speak the Japanese language. The largest ethnic group in Korea is ethnic Koreans, and most people in Mongolia are ethnic Mongolians.

In China, about 92 percent of the people are of the Han ethnic group. The rest of the people belong to 55 different ethnic groups. Han Chinese, spoken by most people, has many dialects. The northern dialect, Mandarin, is China's official language.

Southeast Asia has more ethnic diversity. Ethnic groups in this region include Indonesians, Malays, Burmans, Vietnamese, Laotians, and Thais. Indonesia alone, with its many islands, has about 300 ethnic groups. Hundreds of languages and dialects are spoken throughout the region. Many of the languages were brought by early colonizers.

The people of East Asia and Southeast Asia practice many religions. The major religion in much of the region is Buddhism. In Japan, Buddhism is combined with Shinto, the traditional religion. Other faiths include Confucian, Islam, Hinduism, Daoism, and Christianity. The Communist governments of China and North Korea limit religious practices in those countries.

A variety of arts can be found in East Asia and Southeast Asia. Many paintings reflect a respect for nature, which is part of the Daoism and Shinto religions. Some paintings include poems written in an elegant brush stroke called **calligraphy,** or the art of beautiful writing. Other artists are skilled at weaving, carving, and pottery.

Notes | Read to Learn

People and Cultures (continued)

Illustrating

Based on its description, draw a pagoda in the space below.

Religion and art also are reflected in the architecture of the region. Temples, palaces, and houses are highly decorative. **Pagodas** are temples that are several stories high, with tiled roofs that curve up at the edges.

Literature and theater are popular art forms. Japanese poets write **haiku,** a short poem—usually about nature—that follows a specific structure. Japan also is well-known for its different forms of theater, and Indonesia is famous for its dances.

The family is the center of social life—reflecting the ideas of the ancient Chinese thinker Confucius. He taught that young people should respect their elders and that women should obey their husbands. Education also is highly valued. Many children go to school six days a week. The emphasis on education has helped the region build productive economies.

In cities, people tend to live in small but modern houses. In China, several generations of a family often live together in an apartment. Houses in rural areas are often larger but simpler. In rural Mongolia, people live in circular houses called **yurts** that are made of animal skins. Southeast Asian homes are often built on stilts to avoid flooding.

Section Wrap-Up

Answer these questions to check your understanding of the entire section.

1. **Differentiating** How do East Asia and Southeast Asia differ in terms of ethnic diversity?

   _____
   _____

2. **Determining Cause and Effect** What impact has the importance of education had on the region?

   _____
   _____

Some areas of this region are so densely populated that a lack of privacy is common. On a separate sheet of paper, write a paragraph describing what it might be like to live in such a crowded area.

Chapter 26, Section 2

Chapter 27, Section 1 (Pages 728–733)
China

Big Idea

The characteristics and movement of people impact physical and human systems. As you read, list key facts about China's past and present in the chart below.

	Past	Today
Economy		
Hong Kong and Macao		
Environment		

Notes — Read to Learn

China's Government and Society (pages 729–730)

Listing

List two reasons China's government is criticized.

1. _____
_____
_____

2. _____
_____
_____

The Chinese Communist Party controls China's government. Leaders are not elected by the people. Instead, they gain power by joining the Communist Party and being promoted for their loyalty. The government keeps tight control on political activities and denies individual freedoms. People who criticize the government are punished. In 1989 about 100,000 students and workers protested peacefully for democracy in Beijing's Tiananmen Square. The government sent troops to break up the protest, and thousands were killed or injured.

Other countries have protested China's restriction of **human rights,** or such basic freedoms as freedom of speech and religion. China also is criticized for its actions in Tibet. China took over Tibet in 1950, forcing Tibet's Buddhist leader to leave. He lives in India in exile, meaning he is forced to live someplace other than his own country.

About 65 percent of China's people live in rural areas, usually crowded into river valleys in the east. Village life has improved in recent years. Houses are larger, and many areas now have electricity, radios, and televisions. Motorbikes are becoming more common.

184   Chapter 27, Section 1

China's Government and Society (continued)

China's cities are growing as people migrate to them in search of jobs. Nearly 100 cities in China have populations of more than 1 million people. New office buildings, shopping malls, and apartments are being built quickly.

Economic Changes in China (pages 730–732)

Determining Cause and Effect

Write one cause of air pollution and three causes of water pollution in China.

Air pollution

1. _____

Water pollution

2. _____
3. _____
4. _____

In an attempt to improve the economy, the Chinese government began making free market reforms in the 1970s. People can choose their own jobs, start businesses, and keep their profits. Farmers have some control in the crops they grow and sell.

China now has one of the fastest-growing economies in the world. Only 10 percent of the land can be farmed, yet China is a world leader in growing rice, tea, wheat, and potatoes. Factories make textiles, chemicals, electronic equipment, airplanes, ships, and machinery. The Chinese also have encouraged foreign investment, and many companies are jointly owned by Chinese and foreign businesses. All this growth has increased China's demand for oil and coal.

Hong Kong and Macao are vital to the economic growth in China. Both used to be controlled by European countries, but they came under Chinese control in the late 1990s. They are centers of manufacturing, trade, and finance. Businesses in these two territories help boost the economy of the entire country.

Not everyone in China benefits equally from the economic growth. The standard of living is generally higher in cities than in rural areas. Some people are able to get better jobs, higher wages, and more consumer goods. Others do not make enough money to afford the higher prices of goods, and they have remained poor.

China's environment has been harmed by the country's economic growth. Coal is burned for fuel, leading to air pollution. Factory wastes, sewage, and fertilizers and pesticides have harmed the water. Recently, the government has placed limits on burning coal for fuel.

China's Neighbors (pages 732–733)

Taiwan is an island about 100 miles from China. Most of the people live on the narrow, fertile plain along its western coast. Taiwan used to be a province of China. In 1949 Communists

Chapter 27, Section 1

 **Notes** | **Read to Learn**

China's Neighbors (continued)

Contrasting

How do the economies of Taiwan and Mongolia differ?

_____
_____
_____
_____

took control of China, and the Nationalists fled to Taiwan. Since then, both mainland China and Taiwan claim to rule all of China. Tensions arise whenever Taiwan discusses the possibility of declaring independence. The country has moved toward democracy by allowing political parties other than the Nationalists to form.

Taiwan has a strong industrial economy. Its factories produce textiles, ships, and electronic products. With a highly educated workforce, Taiwan is a center for developing new products.

Landlocked Mongolia is located north of China. Receiving little rain, the landscape includes steppe grasslands and deserts. During the 1200s, Mongolia's people were nomads who used their horse-riding and military skills to create the Mongol Empire. Even today, many of the people raise horses, sheep, goats, cattle, and camels.

Mongolia—nicknamed the "Texas of Asia"—has industries based on herding activities. Some factories use wool to make textiles and clothing, and others use the hides of cattle to make leather and shoes.

In 1990 Mongolia replaced its 70-year-old Communist system with political and economic reforms. Along with these changes, many rural people have moved to cities. About 20 percent of the people now live in Mongolia's capital, Ulaanbaatar.

Section Wrap-Up *Answer these questions to check your understanding of the entire section.*

1. **Naming** What are some recent improvements in rural village life in China?

   _____
   _____

2. **Assessing** Why have some Chinese remained poor despite the growing economy?

   _____
   _____

 **Expository Writing** *On a separate sheet of paper, write a paragraph explaining the changes that are taking place in China's cities.*

186   Chapter 27, Section 1

Chapter 27, Section 2 (Pages 736–741)

Japan

Big Idea

People's actions can change the physical environment. As you read, complete the diagram below. Identify reasons for Japan's successful economy.

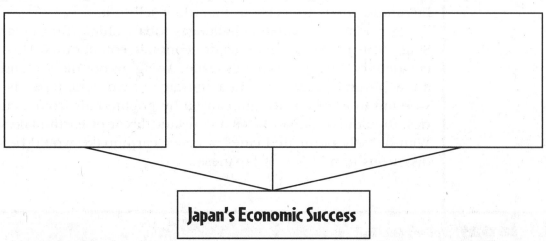

Japan's Economic Success

 **Notes** **Read to Learn**

Government and Economy (pages 737–738)

Describing

Describe Japan's form of government.

_____
_____
_____
_____

The country of Japan includes thousands of islands. The four largest are Hakkaido, Honshu, Shikoku, and Kyushu. Japan's capital, Tokyo, is located on Honshu. Japan is a democratic constitutional monarchy. The emperor is head of state, but elected officials run the government. Voters elect representatives to the national legislature, called the Diet. A prime minister leads the government. Japan's constitution forbids the country from being a military power. The United States keeps military bases there to defend Japan.

Although Japan has few minerals, it is a strong economic power. It imports the raw materials it needs to produce manufactured goods. Because of limited farmland, farmers practice **intensive agriculture,** meaning they grow crops on every available piece of land. Japan grows enough rice for its people, but it imports other food.

An industrial giant, Japan's economy is based on exports. Modern factories and new technologies have made Japan one of the leading producers of steel, cars, ships, cameras, and consumer electronics.

Chapter 27, Section 2

Government and Economy (continued)

Identifying

Identify and underline four challenges facing Japan today.

Japanese businesses work closely with the government. Government-owned banks lend businesses money. The government also passes laws that make it difficult for other countries to sell their products in Japan. As a result, countries like the United States have a **trade deficit** with Japan. This means they buy more goods from Japan than they sell to it.

Japan faces four main challenges today. China, Taiwan, and South Korea have become major economic competitors. This competition threatens Japan's trade. An aging population and a low birthrate may result in a shortage of workers. Japan is working to reduce acid rain caused by polluted air from factories. In addition, Japan faces a constant threat of earthquakes. With densely populated cities, a major earthquake would be disastrous to millions of Japanese.

Life in Japan (pages 739–741)

Stating

How large is the population of Tokyo and its suburbs?

_____

_____

Much of Japan is covered with mountains. Most of the people live in cities on the coastal plains. More than 35 million people live in Tokyo and its surrounding area. The Japanese have adapted to their limited space. They have actually increased their land by depositing dirt near the shoreline to create islands. They also build small houses. In crowded cities, many people live in tall apartment buildings.

In rural areas, people live in farming villages. These villages are connected to nearby cities by high-speed trains and subways. Some people live in villages and commute daily to jobs in cities.

Listing

List the traditional features of Japanese homes and clothing.

_____

_____

_____

_____

Traditional and modern ways are blended in homes, clothing styles, and food. Some Japanese homes are traditional, with wooden floors covered by straw mats called **tatami**. Rooms are used for living space during the day and for sleeping at night. Modern houses have become common, but many have at least one traditional room.

Most Japanese wear Western-style clothes. For special occasions, they wear a long silk robe called a **kimono**, which has an open neck and large sleeves. People eat traditional Japanese meals of rice served with meat, fish, vegetables, and soup. American hamburgers and fried chicken have become popular, however.

Religion

The two major religions in Japan are Shinto and Buddhism. Shinto focuses on respect for nature, love of simple things, and concern for cleanliness and good manners. Buddhism emphasizes

Life in Japan (continued)

Defining

What is anime?

_____
_____
_____

having respect for nature and striving for inner peace. The Japanese blend these two religions, attending Shinto shrines for some events and Buddhist temples for others.

Japanese Pastimes

The pastimes of Japan are a mix of traditional and modern elements. Traditional forms of theater—such as No, Kabuki, and Bunraku—provide entertainment. A more modern activity is going to movies. *Anime,* the Japanese style of animation that began in the late 1900s, is also popular at home and worldwide.

In sports, the Japanese practice the traditional martial arts of judo and kendo. Sumo wrestling remains popular. The Japanese also enjoy the modern sport of baseball.

Section Wrap-Up

Answer these questions to check your understanding of the entire section.

1. **Explaining** Why do many countries have a trade deficit with Japan?

   _____
   _____

2. **Comparing and Contrasting** Compare and contrast Shinto and Buddhism in the Venn diagram below.

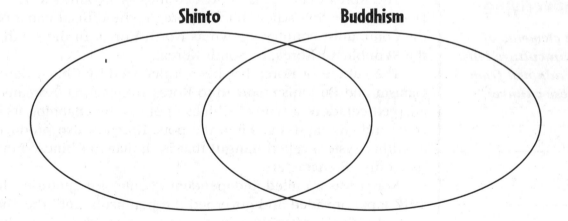

Imagine you are visiting a family in Japan. On a separate sheet of paper, write a letter home describing how daily life there differs from daily life at home.

Chapter 27, Section 2

Chapter 27, Section 3 (Pages 742-746)
The Koreas

Big Idea

Culture groups shape human systems. As you read, compare and contrast the economies of North Korea and South Korea in the Venn diagram below.

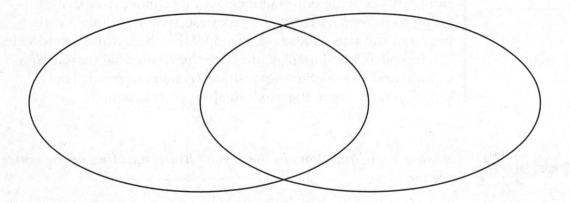

Notes — Read to Learn

South Korea (pages 743–745)

Specifying

What elements of Korean culture have been adapted from Chinese culture?

_____

_____

_____

The Korean Peninsula is divided into two countries. The Democratic People's Republic of Korea is the official name of the Communist country known as North Korea. In the south is the Republic of Korea, or South Korea.

The culture of Korea has been influenced by China. Confucianism and Buddhism spread to Korea from China. Koreans adapted **celadon,** a type of Chinese pottery, by changing its color and cutting designs into the pots. Koreans also produced a writing system called **hangul** that is similar to Chinese but uses only 28 characters.

Korea was a united, independent country for centuries. In 1910 Japan invaded and governed the peninsula until the end of World War II. After the war, Soviet troops took over northern Korea, and U.S. troops moved into the southern half. Korea became divided along the 38th parallel. In 1950 North Korea attacked South Korea to unify the country under Communist rule. This was the beginning of the Korean War. After the war ended in 1953, a **demilitarized zone (DMZ)** separated the two countries. Relations remain tense to this day.

South Korea (continued)

Calculating

What percentage of Koreans are younger than age 30?

_____

Expressing

What does tae kwan do emphasize?

_____
_____

South Korea had to rebuild its economy after the Korean War. It introduced **land reform,** in which large estates were divided into family farms. The government borrowed from foreign banks to build industries. Textiles, iron and steel, cars, ships, and electronic goods were made and then exported. Wages were kept low, and profits were used to repay the loans. Since then, wages have risen, more goods are available to the people, and the standard of living is higher. Today South Korea is a world economic leader.

About 80 percent of South Koreans live in cities. Seoul, the capital, has about 10 million people. Modern styles blend with tradition in the cities' architecture and culture. About 80 to 90 percent of South Koreans own cell phones. The country's population is generally young. More than one-third of the people are younger than 30. People tend to marry at a later age and have few children, so the population is growing slowly.

The major religions in South Korea are Buddhism, Confucianism, and Christianity. Like Japan, Korea has a tradition of martial arts. Tae kwan do, a martial art that emphasizes mental discipline, began in Korea. The people also celebrate traditional holidays. During the fall harvest festival called Chuseok, Koreans leave rice cakes at cemeteries to honor their ancestors.

North Korea (pages 745–746)

Determining Cause and Effect

Give four reasons why North Korea is economically poor.

_____
_____
_____
_____
_____
_____

Since 1948, North Korea has been ruled by Communist dictators. In the late 1940s, Kim Il Sung became North Korea's first ruler. His son, Kim Jong Il, became ruler after Kim Il Sung's death in 1994.

The dictator puts the Communist system above the needs of the people. The government controls all areas of life. People have few freedoms, and traveling into and out of the country is difficult. North Korea tested a nuclear weapon in 2006, which concerns other countries.

North Korea is economically poor. The country has deposits of coal and iron ore, but industries use old equipment and face frequent power outages. The government devotes resources to the military. For many years, it traded with the Soviet Union and Eastern Europe. When communism fell there, North Korea's trade also fell. Investment from other countries is not encouraged.

Poverty is widespread. The government's farm policy has caused hardship. Under this policy, private ownership of land was ended and replaced by large, government-owned farms.

Chapter 27, Section 3

Notes | Read to Learn

North Korea (continued)

Identifying

List products produced in North Korea's factories.

_____

_____

_____

_____

Instead of growing food for themselves, farmers must give their harvests to the government, which distributes food to the people. Farmland is limited, however, and the government farms are not productive. Food must be imported. In some years, the North Korean government refused to take food from world relief organizations. As a result, many people starved. Because of harsh conditions, the infant death rate is high, and many thousands of people have fled the country.

North Korea's government also controls industry. It tells managers and workers what to make. The main products are iron, steel, chemicals, and textiles. North Korea also produces military weapons, some of which are exported.

Section Wrap-Up

Answer these questions to check your understanding of the entire section.

1. **Naming** What are the official names of the two countries on the Korean Peninsula, and when were the countries created?

_____

_____

2. **Explaining** Why is the population of South Korea growing slowly?

_____

_____

Persuasive Writing

In the space provided, write a paragraph explaining why North Korea should abandon its farm policy and allow private family farms.

_____

_____

_____

_____

_____

_____

Chapter 27, Section 4 (Pages 752–758)
Southeast Asia

Big Idea

The physical environment affects how people live. As you read, fill in the chart below with the economic activities of four Southeast Asian countries.

Country	Economic Activities

Read to Learn

Mainland Southeast Asia (pages 753–755)

Summarizing

Complete these sentences.

Myanmar has a

_____

government and a

_____

economy.

Thailand is the only country in Southeast Asia that was never _____.

Malaysia is located on the _____ and _____.

About two-thirds of Myanmar's people are farmers. Most use water buffalo to pull their plows. The main crops are rice, sugarcane, beans, and peanuts. Myanmar exports wood products, gas, rice, and beans. Teakwood is a famous export, but the forests suffer from overharvesting. Myanmar also exports **precious gems**, or valuable stones like rubies, sapphires, and jade. The Irrawaddy River valley is densely populated. Yangon, Myanmar's largest city, is known for its Buddhist temples. Buddhism is the main religion. Since 1948, military leaders have turned Myanmar into a socialist country with a command economy. Aung San Suu Kyi has tried to promote democracy.

Thailand is the only country in Southeast Asia that was never a European colony. Most Thais practice Buddhism and live in rural areas. Many look for jobs in Bangkok, the capital. Tourists come to see Bangkok's beautiful temples and royal palaces. Thailand exports tin, tungsten, precious gems, and rubber. Foreign investors have helped build up industry, and factories produce textiles, clothing, cars, and computer parts.

Malaysia was formed in 1963, when areas on the southern end of the Malay Peninsula and on part of the island of Borneo

Chapter 27, Section 4 193

Mainland Southeast Asia (continued)

Identifying

Around what feature is Cambodia trying to develop a tourist industry?

_____

_____

Naming

What is Vietnam's capital?

_____

Largest city?

_____

united as one country. Exports include raw materials—palm oil, rubber, tin, and woods—as well as manufactured goods—textiles, electronic goods, and cars. Cities such as Kuala Lumpur, the capital, are important trade and industrial centers. Most people are of the Malay ethnic group, and the major religion is Islam.

Singapore lies off the southern tip of the Malay Peninsula. Once covered by rain forests, Singapore today has factories, office buildings, and one of the world's busiest harbors. Because Singapore is a free port, trade is a vital part of the economy. Factories make electronic goods, machinery, chemicals, and paper products. Singapore's people enjoy a high standard of living.

Laos is a poor country on the Southeast Asian mainland. Most people live in rural areas along the fertile Mekong River, where they grow rice and other food. The Communist government discourages religion, but most people are Buddhists.

Cambodia, another poor country, lies south of Laos. In 1975 Communists took control, killed some 2 million people, and left the country in ruins. Cambodia is still recovering. Most people are Buddhists and grow rice. The government is trying to develop a tourist industry centered on Angkor Wat—extraordinary temples of the ancient Khmer Empire.

Vietnam has the largest population in mainland Southeast Asia. Its cities are growing, but most people still live in rural areas. The capital, Hanoi, is located in the north. The largest city is Ho Chi Minh City, named for the first Communist leader. Vietnam's natural resources include coal, petroleum, and metals, all of which are exported. Cash crops—rice, rubber, tobacco, tea, and coffee—are grown along the fertile deltas of the Red and Mekong Rivers. Communist rulers have loosened some controls, and the economy is growing slowly.

Island Southeast Asia (pages 756–758)

Explaining

Why do earthquakes strike Indonesia?

_____

_____

Indonesia, an archipelago of thousands of islands, is located where two tectonic plates meet. As a result, volcanoes erupt and earthquakes occur here. In 2004 about 200,000 Indonesians died when an undersea earthquake launched a huge tsunami. Indonesia is densely populated, with half of its 220 million people living on the island of Java. Jakarta, the capital, is a modern city on Java. Indonesia is rich in oil, natural gas, tin, silver, nickel, copper, bauxite, and gold. Rain forests provide valuable woods. **Mangroves,** or tropical trees that grow along the coasts, help maintain the coastal environment. Trees are being cut at a rapid rate, however.

Island Southeast Asia (continued)

Comparing

Identify common features between countries.

_____ and _____ have volcanic mountains.

_____, _____, and _____ have oil and gas deposits.

Many Indonesians farm. Agricultural exports include rubber trees, oil palms, tea, and coffee. The government is trying to develop tourism, especially on the beautiful island of Bali, which has elaborate Hindu temples. Indonesia is a democracy, but the government finds it difficult to unite the country. Most Indonesians are Muslims, but they are scattered among the islands and belong to different ethnic groups.

Tiny Brunei is located on Borneo. Its economy depends on its large oil and gas reserves. Most food and other goods must be imported.

East Timor lies on the island of Timor. It gained independence from Indonesia in 2002. Most people farm, but the country has oil and natural gas that the government hopes to develop.

The Philippines is made up of about 7,000 islands, many of which have volcanic mountains. Strips of land called **terraced fields** are cut out of the hillsides like stair steps. On these fields, farmers grow rice, sugarcane, bananas, and coconuts for export. Natural resources include gold, iron ore, copper, lead, and zinc. Cities are busy and modern. Factories in Manila, the capital, produce electronics, food products, chemicals, and clothing. About 90 percent of Filipinos are Roman Catholic. The culture blends Malay, Spanish, and American traditions.

Section Wrap-Up *Answer these questions to check your understanding of the entire section.*

1. **Analyzing** What are terraced fields, and why are they used in the Philippines?

   _____
   _____

2. **Identifying** Identify the main religion of four countries in Southeast Asia.

   _____
   _____

Imagine that you work for the tourist bureau of one of the Southeast Asian countries. On a separate sheet of paper, write a paragraph to persuade your reader to visit that country.

Chapter 27, Section 4

Chapter 28, Section 1 (Pages 776–779)
Physical Features

Big Idea

Physical processes shape Earth's surface. As you read, complete the chart below. List one example of each landform in each area, and write a key fact about it.

	Landform	Key Fact
Australia		
New Zealand		
Oceania		
Antarctica		

Notes — Read to Learn

Landforms of the Region (pages 777–778)

Where is the best farmland in Australia and New Zealand located?

Australia

_____

_____

New Zealand

_____

_____

Australia is a country and a continent. It is mostly flat, with few differences in elevation. Narrow plains along the east and southeast make good farmland. The Murray and Darling Rivers run through these plains, and most Australians live there.

The Great Dividing Range rises along Australia's eastern coast. It is an escarpment rather than a true mountain range, because the rocky plateau plunges to lowlands below it. West of the range sprawls a huge area of flat, dry plains and plateaus called the **outback.**

The Great Barrier Reef lies off Australia's northeastern coast. This **coral reef** is a structure formed by the skeletons of small sea animals. It is the largest coral reef in the world, stretching about 1,250 miles.

New Zealand is made up of many islands. The two main islands—North Island and South Island—are separated by the Cook Strait. The country lies along a fault line where two tectonic plates meet. As a result, the North Island has active volcanoes and **geysers,** or hot springs that carry steam and heated water to the Earth's surface. The geysers can erupt as high as 60 feet into the air.

Landforms of the Region (continued)

Listing

List the three types of islands that make up Oceania.

1. _____
2. _____
3. _____

Differentiating

How is an ice shelf different from an ice cap?

_____
_____
_____

The Southern Alps run along the western coast of South Island. Glaciers rest on the mountain slopes. Long ago, they carved steep-sided valleys, called fjords. Today the fjords are filled with sparkling blue water. East of the Alps, the fertile Canterbury Plains form New Zealand's best farming area.

Oceania is a grouping of thousands of islands, including New Zealand, in the Pacific Ocean. Three types of islands—high, low, and continental—can be found in Oceania.

High islands were formed by volcanic activity. These mountainous islands have valleys that fan out into coastal plains. Tahiti and the Fiji Islands are examples of high islands. **Low islands** were formed by coral. Many low islands, such as the Marshall Islands, are **atolls,** or low-lying, ring-shaped islands that surround shallow pools of water. **Continental islands** were formed when tectonic plate movement caused rock to rise and fold from the ocean floor. New Guinea and the Solomon Islands were formed this way. Continental islands have mountains, plateaus, and valleys.

Antarctica is located at the Earth's southern polar region. The Transantarctic Mountains divide Antarctica into two regions. A high plateau rises to the east. The South Pole is located on this plateau. A group of islands linked by ice lies to the west.

Most of Antarctica's highlands and plains are covered by an ice cap. The ice cap spreads into the ocean, where it forms an **ice shelf** above the water. **Icebergs** are huge chunks of ice that break off the ice shelf and float freely in the water.

Natural Resources (page 779)

Specifying

Write down three energy resources of New Zealand.

1. _____
2. _____
3. _____

Natural resources vary throughout the region. Mineral deposits can be found in Australia, including bauxite, copper, nickel, and gold. New Zealand also has deposits of gold, coal, and natural gas. In addition, New Zealand generates hydroelectric power from its rivers and gets geothermal energy from its hot springs.

Few resources are found on the smaller islands of Oceania. Some of the larger islands have oil, gold, nickel, and copper, however. Geologists have discovered that Antarctica has rich deposits of coal and iron ore, which would be difficult and costly to tap because of the harsh climate. Many nations have agreed not to mine Antarctica's mineral wealth in order to preserve the environment.

Chapter 28, Section 1

Natural Resources (continued)

Because of the isolation of the islands in this region, some native plants and animals are unique and do not live anywhere else in the world. Two famous examples are Australia's kangaroos and koalas. These animals are **marsupials,** or mammals that carry their young in a pouch. Another unusual animal is the kiwi, a flightless bird. It is the national symbol of New Zealand.

Section Wrap-Up

Answer these questions to check your understanding of the entire section.

1. **Determining Cause and Effect** How has New Zealand's location on a fault line affected its physical features?
   _____
   _____

2. **Defining** What is an iceberg, and how is it formed?
   _____
   _____

In the space provided, write a description of two or more unique features of this region.

_____
_____
_____
_____
_____
_____
_____

Chapter 28, Section 2 (Pages 782-786)
Climate Regions

Big Idea

Places reflect the relationship between humans and the physical environment. As you read, complete the diagram below. Explain the effects of climate on life in each area.

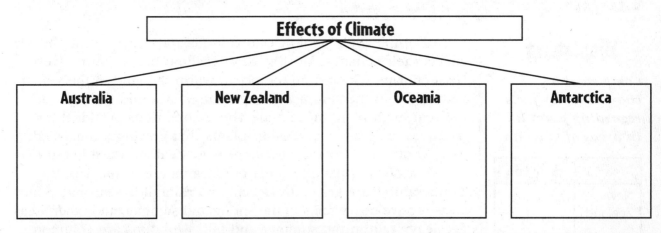

Notes | Read to Learn

Climates of Australia (pages 783-784)

As you read, sketch a simple map of Australia and label its six climate zones.

In general, Australia is a dry continent. Large portions of the outback are covered by deserts. These interior areas receive no more than 8 inches of rain per year. The desert regions are encircled by a steppe climate zone. The steppe receives enough yearly rainfall to allow for some farming. In a dry region west of the Great Dividing Range, wells bring water from a vast underground reservoir called the Great Artesian Basin. This allows people to live in this region even though it is very dry.

Eucalyptus trees can grow in central Australia's desert areas. These trees have thick, leathery leaves that hold in moisture, so they can survive the dry conditions. Other plants have long roots that can reach groundwater during the dry season.

Not all parts of Australia are dry, however. A tropical savanna climate zone covers the far north. Moist, warm air from the ocean rises and cools over this area, bringing monsoon rains. The summers are hot and humid, whereas winters are more pleasant.

A narrow stretch of Australia's northeastern coast experiences a humid subtropical climate. Rainfall is heavy here, and temperatures are warm throughout the year. A marine west coast climate is found along the eastern coast. Summers are warm, winters

Read to Learn

Climates of Australia (continued)

are cool, and rainfall is plentiful. Most of Australia's people live in this area. The southern and western parts of Australia have a Mediterranean climate of warm summers and mild winters.

Climates of Oceania (page 785)

Displaying

Complete this chart with key facts regarding plant life in areas of Oceania.

Area	Plant Life
New Zealand	
High islands	
Low islands	

All of the countries in Oceania are islands. As a result, the sea affects their climates. Most of New Zealand has a marine west coast climate. Winds from the ocean warm the land in the winter and cool it in the summer. Temperatures are mild throughout the year, with plentiful rainfall. These conditions are ideal for **pastures,** or grasses and other plants that grazing animals eat. As a result, many of the people of New Zealand raise livestock.

Most of the smaller islands of Oceania are in the Tropics. Temperatures are generally warm, and rainfall is seasonal. Some areas receive heavy rains in the spring and summer, and other areas get heavy rain in the summer and fall. Typhoons are common. Rainfall also is affected by the elevation of the islands. Islands with higher elevations tend to receive more rain. The mountainous areas of the high islands also tend to have lower temperatures.

Vegetation in Oceania varies based on the type of island. Because high islands have more rainfall as well as fertile, volcanic soil, they have a greater variety of plants. The conditions on low islands can support only a few kinds of plants, such as coconut palms and breadfruit trees. **Breadfruit,** a starchy pod that can be cooked in a number of different ways, is a food staple in Oceania.

The Climate of Antarctica (page 786)

Explaining

Give three reasons why Antarctica is so cold.

1. _____
2. _____
3. _____

Antarctica never receives direct sunlight. Therefore, the temperatures are extremely cold, ranging from a high of –4°F in the summer to –129°F in the winter. In addition, harsh winds blow across the region. The only people who live in Antarctica are scientists who stay there for brief periods of time while doing research.

The air of Antarctica is so cold that it is unable to hold much moisture. Because humid air is needed to trap the sun's warmth, Antarctica's dry air contributes to its coldness. Even though Antarctica is covered by ice, it is actually a desert.

Notes | Read to Learn

The Climate of Antarctica (continued)

People do not live in Antarctica, but some plants and animals do. Penguins and other marine mammals feed off of the rich sea life. Tiny, sturdy plants called **lichens** grow in rocky areas along the coasts.

Section Wrap-Up
Answer these questions to check your understanding of the entire section.

1. **Naming** What kinds of trees grow in central Australia and why?
   _____
   _____

2. **Comparing and Contrasting** Complete this Venn diagram with key facts about high islands and low islands in Oceania.

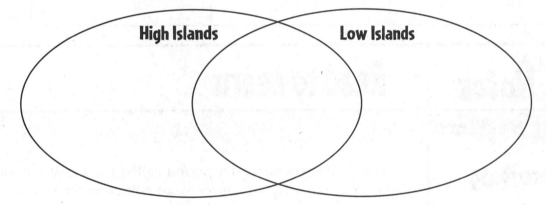

Descriptive Writing
In the space provided, write a paragraph describing the human, plant, and animal life in Antarctica.

_____
_____
_____
_____
_____

Chapter 28, Section 2

Chapter 29, Section 1 (Pages 792-797)
History and Governments

Big Idea

Geographic factors affect where people settle. As you read, complete the chart below by organizing key facts about the first peoples to settle each area.

First Settlers		
Australia	**New Zealand**	**Oceania**

Read to Learn

First Settlers (page 793)

Describing

Write a sentence describing the Ice Age.

_____
_____
_____
_____
_____

The Earth experienced a period called the Ice Age about 40,000 years ago. Temperatures were colder, much of the Earth's water was frozen, and ocean levels were lower than they are today. During this time, people from Southeast Asia were able to travel to Oceania and Australia by land or by canoe. When the Ice Age ended and ocean levels rose, these settlers were cut off from the rest of the world.

The descendants of these first Australians are called Aborigines. Early Aborigines traveled in small family groups as they hunted, gathered plants, and searched for water. Aborigines made a weapon called a **boomerang**, which is a flat, bent piece of wood that hunters throw to stun their prey. If the boomerang does not hit the prey, it circles back to the person who threw it.

Ancient rock paintings tell much about the Aborigines' early history. Aborigines believed that a powerful spirit created the land and that their role was to care for the land.

By 1500 B.C., the settlers of New Guinea and nearby islands had built large canoes that could travel long distances. Over time, they reached remote islands such as Fiji, Tonga, and Hawaii.

First Settlers (continued)

The Maori people left the Pacific island region of Polynesia between A.D. 950 and 1150 and settled in New Zealand. They built villages, hunted, fished, and farmed on North Island and South Island. Farmers grew root crops. The Maori people also developed skills in wood carving.

The European Era (pages 794–795)

Specifying

Who were the first British settlers of Australia?

_____

Stating

Write two reasons why foreign countries were attracted to the Pacific Islands.

1. _____

2. _____

Europeans explored the South Pacific region from the 1500s through the 1800s. In the late 1700s, Captain James Cook claimed eastern Australia for Britain. At that time, the prisons in Britain were overcrowded; thus, Australia was established as a colony where convicts could serve their sentences. By the mid-1800s, Britain stopped sending convicts to Australia, but many free British people continued to migrate there. Many settlers were wheat farmers and sheep ranchers. Gold was discovered in 1851, which led to a new rush of settlers.

At first, relations between the British and the Aborigines were peaceful. As the settlers took more land for ranching and farming, however, the Aborigines fought to defend their traditional hunting grounds. Warfare and disease reduced the Aborigine population to about 80,000 by the late 1800s.

British settlers arrived in New Zealand in the early 1800s. In 1840 the Maori chiefs and Britain signed the Treaty of Waitangi—the Maori agreed to British rule but were allowed to keep their land. Nevertheless, British farmers and sheep ranchers continued to move onto Maori land, and war broke out in the 1860s. The Maori lost both the war and their land.

As global trade grew, the United States and European countries became interested in the Pacific Islands as trading ports and refueling stations. They began to set up colonies. By the early 1900s, most of the Pacific area was under foreign control.

Independent Nations (pages 796–797)

Australia gained independence from Britain in 1901. New Zealand became independent in 1907. Both countries are parliamentary democracies with elected representatives who choose a prime minister to head the government. Australia has a federal government, with power divided between the national and state or territorial governments. In New Zealand, the national

Notes | Read to Learn

Independent Nations (continued)

Sequencing

Identify, in order, independence dates in the region.

1. _____
2. _____
3. _____
4. _____

government holds all major power. In 1893 New Zealand became the first country to give women the right to vote. It was also among the first countries to provide government help to the elderly, the sick, and the jobless. Even so, the Maori in New Zealand and the Aborigines in Australia suffered from discrimination.

Most of Oceania remained under foreign control during the early 1900s. Japan won control of Germany's Pacific colonies after World War I. During World War II, the United States and Japan battled in the Pacific islands. After Japan's defeat, its Pacific colonies became trust territories of the United States. **A trust territory** is an area that is temporarily placed in the control of another country. Since the 1960s, most of these trust territories have become independent.

Antarctica's status also has gone through changes in the 1900s. Several countries claimed land there. In 1959 many countries signed the Antarctic Treaty. These countries now share Antarctica for peaceful scientific research.

Section Wrap-Up

Answer these questions to check your understanding of the entire section.

1. **Describing** How did relations between the British settlers and the Aborigines change over time?

   _____
   _____

2. **Comparing and Contrasting** Use this Venn diagram to compare and contrast the governments of Australia and New Zealand.

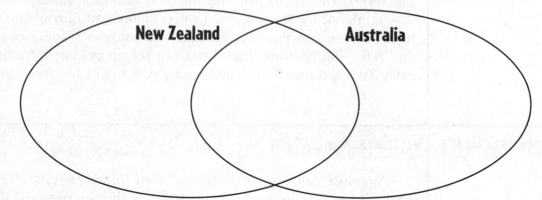

On a separate sheet of paper, write a paragraph explaining how the status of the Pacific islands changed in the 1900s.

Chapter 29, Section 2 (Pages 799–804)
Cultures and Lifestyles

Big Idea

Culture groups shape human systems. As you read, complete the Venn diagram below. Compare and contrast the people of Australia and New Zealand with the people of Oceania.

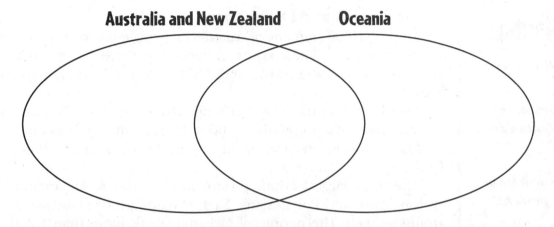

Australia and New Zealand   **Oceania**

Read to Learn

The People (pages 800–801)

Defining

As you read, explain the terms *bush* and *station* by using each in a sentence.

_____
_____
_____
_____
_____

Climate, landforms, and culture contribute to population differences throughout the region. Australia has more than 20 million people. The population of New Zealand is about 4 million. Low birthrates have slowed the population growth rate in both countries, but immigration has increased. Most of Australia's people live in coastal areas that have a mild climate, fertile soil, and access to the ocean. Most New Zealanders also live along the coasts.

In Australia and New Zealand, more than 85 percent of the people live in urban areas. Sydney and Melbourne, Australia, and Wellington and Auckland, New Zealand, are thriving cities. A small number of Australians live in rural areas called the **bush.** Some of these people work on cattle and sheep ranches, or **stations.** Others farm or work in mining camps.

Populations are growing rapidly on other islands of Oceania. Papua New Guinea is the most populous country, with about 6 million people. Some of the smaller islands have problems with overcrowding. Tiny Nauru has only 10,000 people, but they are crowded into 9 square miles. Many Pacific Islanders have begun migrating to other parts of the world.

The People (continued)

Identifying

What is the largest city in Oceania?

_____

Complete this sentence.

On average, out of 200 New Zealanders,

_____

of them would be of European descent.

Oceania has few large cities. Port Moresby, the capital of Papua New Guinea, is the largest urban area. Most Pacific Islanders live in rural villages.

The climate and terrain of Antarctica are not favorable for people to live there. The only residents are scientists and tourists who remain for short visits.

Ethnic Groups and Languages

More than 90 percent of Australia's people are of British and Irish descent. The country is becoming more diverse as the Aborigine population grows and additional immigrants arrive from Asia.

People of European backgrounds make up about 75 percent of New Zealand's population. About 13 percent of the people are Maori, and Asians and Pacific Islanders make up another 10 percent.

The three largest ethnic groups in Oceania are Melanesians, Micronesians, and Polynesians. Each is made up of many smaller groups as well. The people of Oceania speak more than 1,200 languages. Papua New Guinea alone has 700 languages. Many people there speak a **pidgin language,** which is formed by combining several languages.

Culture and Daily Life (pages 803–804)

Write a sentence expressing the main idea of this passage.

_____
_____
_____
_____

The cultures of the region have been influenced by European and Pacific traditions. Christianity, brought by Europeans during the 1700s and 1800s, is the most widely practiced religion. Traditional religions are still practiced in some areas, however. The Aborigines, for example, believe in Dreamtime, a time long ago when wandering spirits created the world. Aborigines believe all natural things have a spirit and are connected to one another.

The arts in Australia tie together past and present. Aborigines made paintings on rocks to show the relationship between people and nature. Painters of European descent feature Australian landscapes in their works. Writers and filmmakers use local themes.

New Zealand's Maori artisans are experts in canoe making, weaving, and wood carving. Storytellers share the people's history and myths. **Action songs,** developed in the 1900s, blend traditional dance with modern music.

Notes | Read to Learn

Culture and Daily Life (continued)

Speculating

What is poi? What Western food do you think it tastes like?

_____
_____
_____
_____

In Oceania, dances play a key role in important events. Pacific Islanders pass on their culture through storytelling. Movements of the dancers are often used to tell these stories.

Daily life also blends tradition with modern ways. People of European backgrounds tend to live in nuclear families. Aborigines, Maori, and Pacific Islanders tend to live with extended families. Maori households may include three or four generations.

Houses in Australia are often one floor and made of brick or wood with tiled roofs. In New Zealand, people live in timber houses or in stone cottages. City residents live in Western-style apartments or houses. In Oceania, many homes, like the Samoan *fale,* have open sides that allow ocean breezes to circulate.

Meat is a staple of the diet in Australia and New Zealand. In Oceania, people eat fish, pork, yams, breadfruit, and taro. One regional food is **poi,** a paste made by mashing the fleshy bulb of the taro plant.

Outdoor sports are popular. People swim, surf, and scuba dive. Boat racing is a favorite in Oceania. Rugby and cricket, brought by the British, are popular in Australia and New Zealand.

Section Wrap-Up

Answer these questions to check your understanding of the entire section.

1. **Identifying** Where do most Australians and New Zealanders live? Why?

   _____
   _____

2. **Listing** Complete this table with several examples of European influence and traditional culture in the region.

European Influence	Traditional Culture

On a separate sheet of paper, create an advertisement to attract tourists to a specific area of this region—Australia, New Zealand, Oceania, or Antarctica.

Chapter 30, Section 1 (Pages 812–816)
Australia and New Zealand

Big Idea

People's actions can change the physical environment. As you read, compare the economies of Australia and New Zealand in the Venn diagram below.

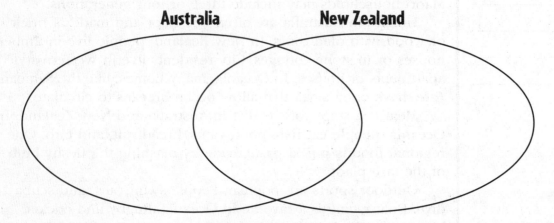

Read to Learn

Australia (pages 813–814)

Stating

Write down five areas from which immigrants to Australia originated.

1. _____
2. _____
3. _____
4. _____
5. _____

Australia is large in land area, but it has only 20.6 million people. More skilled workers are needed to help build the economy, so the government has encouraged immigration. Initially, most immigrants were from the British Isles. Since the 1970s, immigrants have come from Asia, South Africa, Latin America, and Oceania.

Aborigines, the original settlers of Australia, have experienced discrimination. Although the government is working to improve their conditions, Aborigines still suffer from poverty and poor health care. In the 1980s, a group of Aborigines filed a **lawsuit,** or legal action in court, to prevent mining on land they said belonged to them. The court ruled in their favor. Other Australians worry that they might lose land to the Aborigines. The government is trying to balance the claims of both groups.

Australia has a prosperous economy. It is based partly on the export of resources, including iron ore, nickel, zinc, bauxite, gold, diamonds, coal, oil, and natural gas. China and Japan purchase many of these resources. The dry climate limits farming, but irrigation allows farmers to grow grains, sugarcane, cotton, fruits, and vegetables. The main agricultural activity is sheep

Notes | Read to Learn

Australia (continued)

Listing

List three environmental challenges of Australia.

_____
_____
_____

and cattle herding. Australia exports wool, lamb, beef, and cattle hides. Some ranchers raise **merinos,** a breed of sheep known for its fine wool. Most of Australia's industries are located near Sydney and Melbourne. Factories make food products, transportation equipment, cloth, chemicals, and high-technology goods. Tourism plays a large role in the economy as well.

Australia faces environmental challenges. Deforestation and overgrazing have ruined much of the topsoil. In addition, animals brought by settlers from other countries have threatened some of Australia's native wildlife. For example, Hawaiian cane toads were brought to Australia to eat insects that damaged sugarcane. The toads did not eat the insects, but their poisonous skin kills other animals. Australians continue to debate how to solve their environmental problems without hurting the economy.

New Zealand (pages 815–816)

Applying

Why is the ethnic balance of New Zealand expected to change?

_____
_____
_____
_____

Identifying

What provides electricity for New Zealanders?

_____
_____

Like Australia, people in New Zealand are of European origin. The Maori were the original settlers, and they make up the largest non-European group. In 1840 Maori leaders signed the Treaty of Waitangi with Great Britain, which granted the Maori certain land rights. The Maori today use that treaty to reclaim land they feel has been taken unfairly. The Maori have won a number of lawsuits. As in Australia, New Zealanders of European descent fear they will lose land.

New Zealand's population is changing. Pacific Islanders in search of work are immigrating to New Zealand. People from East Asia and Southeast Asia also are moving to the country. The population growth rate of these new immigrant groups is high, as is the growth rate of the Maori. The growth rate among whites is low, however, so the ethnic balance of New Zealand probably will change in the future.

Sheep form the backbone of New Zealand's economy. Sheep-based exports include wool and meat. The cattle industry produces butter, cheese, and meat exports. Other economic goods include wood and paper products from forests along the country's mountains. Farming and wine making are thriving businesses. Major crops are apples, grapes, barley, wheat, and corn. Another important crop is **kiwifruit,** a small oval fruit with a brownish-green, fuzzy skin.

Resources mined in New Zealand include gold, coal, and natural gas. Hydropower and geothermal energy are used to

Chapter 30, Section 1

Notes | Read to Learn

New Zealand (continued)

Explaining

Why is trade so important to New Zealand?

_____
_____
_____

produce electricity. Factories turn out fertilizer, shoes, machinery, and vehicles. Service industries and tourism also are major sources of income for the country.

New Zealand is relatively small. Therefore, the economy depends in large part on trade with other countries. In the past, New Zealand's two major trade partners were the United Kingdom and Australia. Australia is still a key trading partner, but the United Kingdom has lessened in importance. Other vital trade partners today include the United States and countries in East Asia.

Section Wrap-Up

Answer these questions to check your understanding of the entire section.

1. **Comparing** What are some similarities between the Aborigines of Australia and the Maori of New Zealand?

   _____
   _____
   _____

2. **Explaining** How have New Zealand's trading partners changed?

   _____
   _____
   _____

In the space provided, write a letter to the editor of a newspaper encouraging unemployed workers in other countries to immigrate to Australia.

_____
_____
_____
_____
_____
_____
_____

Chapter 30, Section 2 (Pages 822–826)

Oceania

Big Idea

Patterns of economic activities result in global interdependence. As you read, complete the diagram below by writing three important facts about the economies of Oceania. In the large box, write a generalization that you can draw from those facts.

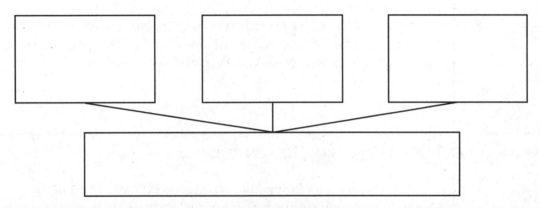

 **Read to Learn**

Melanesia (pages 823–824)

Listing

List four products that are grown on plantations in Papua New Guinea.

1. _____
2. _____
3. _____
4. _____

Geographers organize Oceania into three subregions—Melanesia, Micronesia, and Polynesia. The islands of Melanesia are located across the Coral Sea from Australia. Papua New Guinea is the largest and most-populous country in Melanesia. Its people belong to different Papuan or Melanesian ethnic groups and speak more than 700 languages. Many people are subsistence farmers or work on plantations that grow and export coffee, oil palm trees, cacao trees, and coconut palms. Coconut oil from **copra,** the meat from dried coconuts, is used to make margarine, soap, and many other products. The cash crops grown on plantations do not feed the people, so food must be imported. Papua New Guinea supports its economy by extracting deposits of oil, gold, copper, silver, iron, and zinc from land and the ocean floor.

The people of the Fiji Islands are ethnic Melanesians and South Asians—descendants of British Indians who came to Fiji to work on plantations in the 1800s. These two groups struggle to control the government, a conflict that has harmed Fiji's economy. Foreign companies are afraid to invest in the country, and tourists stay away.

 **Read to Learn**

Melanesia (continued)

Stating

What is the lingua franca of Vanuatu?

People in the Solomon Islands live by subsistence farming and fishing. They tend to follow traditional ways, such as using shells or feathers as money. On the volcanic island of Vanuatu, tradition is strong as well. People there honor volcanic spirits in their religious ceremonies. More than 100 languages are spoken in Vanuatu. Many people use Bislama as their **lingua franca,** or common language, for communication and trade. Tourism is growing there.

New Caledonia, a French territory in Melanesia, exports its rich deposits of nickel. About one-third of the people are of French origin. Some New Caledonians want independence from France, however.

Micronesia and Polynesia (pages 824–826)

Naming

Name two U.S. territories in Micronesia.

Defining

As you read, explain what fa'a Samoa is by using the term in a sentence.

Micronesia and Polynesia are scattered over the Pacific Ocean. They include high volcanic islands as well as low, ring-shaped atolls.

Japan and the United States fought a number of battles on the Micronesian islands during World War II. Many islands have become independent since the 1970s, but American influence remains strong. The United States pays a fee to keep military bases on some islands, and the bases also provide jobs to islanders. The Northern Mariana Islands and Guam are U.S. territories.

People on the volcanic high islands live as subsistence farmers. They grow yams, sweet potatoes, and cassava in the fertile soil. People on the low islands fish and grow breadfruit, taro, and bananas. The poor soil cannot support many crops, so most food is imported.

Several Micronesian islands have phosphates, a mineral salt that makes fertilizer. The Federated States of Micronesia and the Marshall Islands lack the money to mine their phosphates. The deposits on Kiribati are gone now, and they are almost depleted on Nauru. These countries now rely on foreign aid from Japan, the European Union, and Australia. Nauru's government is investing overseas and trying to develop service industries.

Polynesia's islands are located southeast of Micronesia. Some of the islands are independent, such as Samoa and Tonga. Others, such as French Polynesia, remain under European control. Many Polynesians are subsistence farmers. Several of the islands rely completely on foreign aid to survive. Samoa and Tonga have strong tourist industries and also export timber.

Micronesia and Polynesia (continued)

Identifying

Identify two activities that have harmed the environment in Micronesia and Polynesia.

1. _____
2. _____

Other Polynesian industries include canning tuna and issuing postage stamps for collectors.

Samoans call their way of life the *fa'a Samoa.* This lifestyle stresses living in harmony with the community and the land. Samoans are famous for their music, dance, handicrafts, and tattoos.

Environmental Issues

The people and environments in Oceania have been harmed by human activities. In the late 1940s, the United States and other countries carried out nuclear weapons testing in the Pacific, unaware of its dangers. People were exposed to radiation that caused illnesses and death. The radiation also poisoned the land, water, and vegetation. Although the testing has stopped, its effects are still felt by the people and the environment. The United States has given millions of dollars to help Marshall Islanders who were affected. In 1987 New Zealand announced that nuclear-powered ships could not enter its waters. France planned but canceled nuclear testing in the 1990s.

Phosphate mining also has damaged the environment. About 80 percent of Nauru cannot support human life. Native birds are threatened by the loss of their **habitats,** or natural surroundings. Nauru is seeking foreign aid to restore its land.

Section Wrap-Up *Answer these questions to check your understanding of the entire section.*

1. **Determining Cause and Effect** How has the conflict between Melanesians and South Asians affected Fiji's economy?

   _____
   _____

2. **Analyzing** How does the United States contribute to the economy of some Micronesian islands?

   _____
   _____

On a separate sheet of paper, write a paragraph explaining the impact of phosphate mining on the economy and environment in Micronesia.

Chapter 30, Section 3 (Pages 827–830)
Antarctica

Big Idea

All living things are dependent upon one another and their surroundings for survival. As you read, complete an outline of the section on the lines below.

I. First Main Heading _____
   A. Key Fact 1 _____
   B. Key Fact 2 _____
II. Second Main Heading _____
   A. Key Fact 1 _____
   B. Key Fact 2 _____

Read to Learn

International Cooperation (pages 828–829)

Specifying

What did the Antarctic Treaty state could NOT take place in Antarctica?

_____
_____
_____
_____
_____

Antarctica was first sighted in the 1820s. Scientists and seal hunters visited coastal areas of the continent, but they rarely ventured into the interior. In 1911 explorers reached the South Pole. After that achievement, explorers wanted to see what the rest of the continent had to offer.

International Agreements

Some countries, eager to find mineral deposits, made claims to territory in Antarctica. Other countries, including the United States, disagreed with these claims. In the 1950s, several countries began to work together on scientific research in Antarctica.

In 1959 twelve countries signed the Antarctic Treaty. The purpose of this agreement was to prevent future conflicts over territorial claims. The treaty established that Antarctica would be used only for peaceful, scientific purposes. Antarctica could not be used for any military purposes, including weapons testing. A total of 45 countries have signed the Antarctic Treaty. It has been extended to prevent mining in Antarctica and to protect its environment.

International Cooperation (continued)

Naming

Name two types of scientists who study Antarctica.

1. _____
2. _____

Scientific Research

Antarctica is home to scientific research stations from many countries. During January—summer in Antarctica—thousands of scientists come to study the continent's land, plants, animals, and ice.

Scientists conduct a variety of research. Geologists have found the remains of trees millions of years old. These scientists believe that Antarctica was once joined to Africa and South America. Climatologists study samples of ice from deep beneath the surface to learn about the climate from thousands of years ago.

Antarctica's Environment (pages 829–830)

Listing

List the animals that were threatened with extinction.

1. _____
2. _____

Illustrating

Show how ice loss could affect an entire chain of life in Antarctica.

Loss of Ice
_____
↓
_____
↓
_____
↓
_____

Antarctica has a harsh yet fragile environment. The seas around Antarctica are home to penguins, seals, fish, whales, and many kinds of flying birds. Whales and seals once were hunted almost to the point of **extinction,** or disappearance from the Earth. Many countries have taken action to protect these animals.

Environmental Challenges

Although countries have agreed to protect Antarctica's environment, the continent is affected by human activity elsewhere. Humans may be contributing to global warming. Higher temperatures may lead to loss of ice in and near Antarctica. If this happens, the plants that live on the ice also will disappear. These plants are eaten by **krill,** tiny shrimplike creatures. Krill, in turn, are the main food source for many of the larger animals in Antarctica. The survival of the entire chain depends on the ice.

Areas beyond Antarctica also would be affected by an ice melt in the region. Scientists believe that such a melting would raise sea levels and possibly lead to flooding of the low islands in Oceania and of crowded coastal cities.

Antarctic research showed another environmental challenge—the weakening of the ozone layer. **Ozone** is a gas that forms a protective layer around the Earth in the atmosphere. It protects all life on Earth from certain harmful rays of the sun. Scientists first noticed a "hole" in the ozone layer above Antarctica in the 1980s. They believe that the loss of ozone could lead to higher rates of skin cancer and might contribute to global warming.

Chapter 30, Section 3

Antarctica's Environment (continued)

Classifying

What type of product harms ozone?

_____

Many countries have taken steps to protect the ozone layer. Some have limited the use of aerosol sprays and other products that are thought to cause ozone loss. These products contain chemicals that can collect in the atmosphere. When the sun's rays reach these chemicals, they form new chemicals that damage the ozone.

Section Wrap-Up

Answer these questions to check your understanding of the entire section.

1. **Sequencing** Complete the time line below by identifying key events and dates about Antarctica's history.

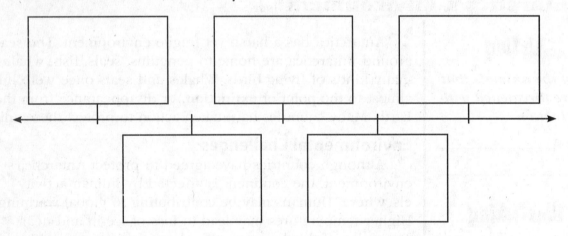

2. **Making Connections** What is the ozone layer, and what is its connection to global warming?

_____

_____

Descriptive Writing

Imagine you are a scientist working in Antarctica. On a separate sheet of paper, write a paragraph describing your experience.